LIMITS OF LIFE

LIMITS OF LIFE

*Proceedings of the Fourth College Park Colloquium
on Chemical Evolution,
University of Maryland, College Park, Maryland, U.S.A.,
October 18th to 20th, 1978*

Edited by

CYRIL PONNAMPERUMA

*Laboratory of Chemical Evolution,
Dept. of Chemistry, University of Maryland, College Park, U.S.A.*

and

LYNN MARGULIS

Dept. of Biology, Boston University, Boston, U.S.A.

D. REIDEL PUBLISHING COMPANY

DORDRECHT : HOLLAND / BOSTON : U.S.A.
LONDON : ENGLAND

Library of Congress Cataloging in Publication Data (Revised)

College Park Colloquium on Chemical Evolution, 4th, 1978.
   Limits of Life.

   Includes index.
   1. Life (Biology)–Congresses. 2. Adaptation (Biology)–Congresses.
3. Chemical evolution–Congresses. I. Ponnamperuma, Cyril, 1923–
II. Margulis, Lynn, 1938–   III. Title.
QH501.C64 1978   577   80-19612
ISBN 90-277-1155-0

Published by D. Reidel Publishing Company,
P.O. Box 17, 3300 AA Dordrecht, Holland.

Sold and distributed in the U.S.A. and Canada
by Kluwer Boston Inc.,
190 Old Derby Street, Hingham, MA 02043, U.S.A.

In all other countries, sold and distributed
by Kluwer Academic Publishers Group,
P.O. Box 322, 3300 AH Dordrecht, Holland.

D. Reidel Publishing Company is a member of the Kluwer Group.

All Rights Reserved
Copyright © 1980 by D. Reidel Publishing Company, Dordrecht, Holland
No part of the material protected by this copyright notice may be reproduced or
utilized in any form or by any means, electronic or mechanical,
including photocopying, recording or by any informational storage and
retrieval system, without written permission from the copyright owner.

Printed in The Netherlands

TABLE OF CONTENTS

PREFACE	vii
EDITORS' INTRODUCTION	ix
PETER MAZUR / Limits to Life at Low Temperatures and at Reduced Water Contents and Water Activities	1
RICHARD Y. MORITA / Biological Limits of Temperature and Pressure	25
E. IMRE FRIEDMANN / Endolithic Microbial Life in Hot and Cold Deserts	33
HORDUR KRISTJANSSON and CYRIL PONNAMPERUMA / Purification and Properties of Malate Dehydrogenase from the Extreme Thermophile *Bacillus Caldolyticus*	47
JAMES T. STALEY / The Gas Vacuole: An Early Organelle of Prokaryote Motility?	55
JANOS K. LANYI / Physical Chemistry and Evolution of Salt Tolerance in Halobacteria	61
STJEPKO GOLUBIC / Halophily and Halotolerance in Cyanophytes	69
S. E. CAMPBELL / Soil Stabilization by a Prokaryotic Desert Crust: Implications for Precambrian Land Biota	85
MICHAEL J. NEWMAN / The Evolution of the Solar 'Constant'	99
JOEL S. LEVINE, ROBERT E. BOUGHNER, and KATHRYN A. SMITH / Ozone, Ultraviolet Flux and Temperature of the Paleoatmosphere	105
JAMES C. G. WALKER / Atmospheric Constraints on the Evolution of Metabolism	121
ANDREW H. KNOLL / Archean Photoautotrophy: Some Alternatives and Limits	133
H. G. THODE / Sulphur Isotope Ratios in Late and Early Precambrian Sediments and their Implications Regarding Early Environments and Early Life	149
MANFRED SCHIDLOWSKI / Antiquity and Evolutionary Status of Bacterial Sulfate Reduction: Sulfur Isotope Evidence	159
KENNETH H. NEALSON and BRADLEY TEBO / Structural Features of Manganese Precipitating Bacteria	173
WILLIAM A. BONNER, NEAL E. BLAIR, and RICHARD M. LEMMON / The Radioracemization of Amino Acids by Ionizing Radiation: Geochemical and Cosmochemical Implications	183
INDEX	195

PREFACE

This volume is the fourth in the series of the Proceedings of the College Park Colloquia on Chemical Evolution. These Colloquia, and the resulting Proceedings, are presented in the interest of fostering the impact of the interdisciplinary nature of chemical evolution on contemporary scientific thought.

EDITORS' INTRODUCTION

The Fourth College Park Colloquium on Chemical Evolution was held on October 18 - 20, 1978 at the University of Maryland. The meeting, supported by the National Aeronautics and Space Administration and the National Science Foundation, centered on the variable environments, both past and present, in which living organisms have survived, grown, and evolved - the limits of life. Previous colloquia had emphasized the Giant Planets (1974)[1], Early Life during the Precambrian (1975)[2] and Comparative Planetology (1976)[3]. The College Park Colloquia have been noted for the broad interdisciplinary nature of the training and interests of the participants. The fourth meeting was no exception with the participation of approximately 85 researchers, representing many academic fields. As with previous meetings, the interdisciplinary approach to the question of the limits of life encouraged the exchange of knowledge and information.

A major scientific aspiration is to understand why living systems are restricted to certain environments. Could life exist on the surfaces of stars, extraterrestrial planets, meteorites, or interstellar particles? Are certain elements, in proper proportions and chemical form, essential for growth and reproduction? How do extreme temperatures, salt concentrations, pressures and desication prevent life from thriving? On Earth, different species have evolved the mechanisms to thrive in conditions lethal to others. What are these mechanisms and what are the ranges of adaption to toleration and growth under extreme environments? Molecular oxygen, ultraviolet and ionizing as well as visible radiation, extreme acidities and alkalinities also limit life, but perhaps not absolutely. An astounding range of tolerance to these environmental vagaries that tends to be species specific has been reported.

It is a striking observation that both the abundance of individuals and the diversity of species increase at interfaces where solid surfaces meet water and air. Most species complete their cycles in the liquid phase of oceans and lakes; a few complete their life cycle on dry land. Yet no species seems able to exist for its entire life cycle suspended in the gas phase. What then limits life on various substrata? What precludes life from invading the atmosphere, and why have so many organisms developed preferences for solid substrata often of a very specialized nature?

This conference publication could be valuable indeed if it were able to face all of these questions and systematically answer them. Unfortunately, present research allows us to ask more questions than we can answer. The task has barely begun. Why life, in general, and certain species, in particular, show varying sensitivities to ambient changes is a research problem of high order, and one in which only idiosyncratic forays have been undertaken. These proceedings reflect the rather random nature of the research itself. Furthermore, they reflect a consistent truth that confronts all investigators dealing with real life in the natural world: environmental extremes are not isolated, and their changes are neither predictable nor uniform with time.

High salt environments, for example, may also have extreme temperatures or limited visible radiation. Low temperature environments may be subject to high hydrostatic pressures as well. The researcher can rarely vary all the critical environmental factors in a regular fashion and isolate them for study. A natural historian interested in studying extreme environments in the field is fortunate if he can identify even the major species components of the community with which he works.

During the Colloquium, several astronomical and cosmological talks were presented. M. Newman's discussion of the changes in solar luminosity, its possible mechanism and effects on the Earth and the presentation of W. A. Bonner et al. on evidence for radioracemization of amino acids appear in this publication. A pioneering attempt at simultaneously understanding the nature of the early Archean atmosphere and its interaction with the evolving microbiota was made by J. C. G. Walker. This effort was enhanced by a discussion of atmospheric ozone at a time, presumably during the Proterozoic Aeon, when the concentration of atmospheric oxygen was less than it is at present. The relationship between ozone, ultraviolet light flux, biogenically modulated oxygen, nitrogen, hydrogen, and the atmospheric temperatures in a 2×10^9 year-old troposphere were modeled by J. S. Levine, R. E. Boughner, and K. A. Smith.

Perhaps because they represent different scientific disciplines the Colloquium participants have recognized that the continual exchange of information and interaction is a prerequisite to understanding the complexities of the biota and its effects on the atmosphere. Such joining of forces and mutual comprehension must be achieved to attain an understanding of extant life as well as of the past life. In this spirit, A. Knoll's paper reconstructed the worldwide geological regimes that prevailed in the Archean (3.5 - 2.0×10^9 years ago) and showed how these differ from those of the stromatolite and microfossil-rich Proterozoic (2.0 until 0.6×10^9 years ago) sediments. Both H. G. Thode and M. Schidlowski reconstructed some environments of the past. They suggested that the ratios between the stable isotopes of sulfur (δS^{34}) can be used to infer the nature and extent of microbial communities in ancient basins. S. Campbell illustrated the manner cyanophyte dominated mats formed a desert crust and related this to conditions for Precambrian land biota.

Some contributions, such as that by E. I. Friedmann, describing the microbial communities present in the very hot deserts of the tropics and the extremely cold ones of the Antarctic, dealt with field studies of live organisms and contained many surprising observations. Neither the community composition nor the mechanisms of coping with the multiplicity of extreme conditions are the same in the two cases. S. Golubic, in another example, pointed out that both the magnitude and the rate of variation of the extremity must be taken into account. Using examples taken from studies on cyanobacteria (blue-green algae), he pointed out the importance of knowing not only if the organism is growing or merely surviving in saline conditions, but also the rate of change of salt concentration over time. J. Lanyi also dealt with adaptations to high salt, and he reported on laboratory studies concerned with biochemistry of adaptation to extremely high salinities. He showed how there are many correlated mechanisms for growth at high sodium chloride concentrations that involve energy expenditure as well as dramatic alterations in the proteins of the salt-requiring bacteria.

P. Mazur presented an analysis of one of the most definitive limits to life: liquid water. In a thorough review he was able to show the physico-chemical mechanisms involved in the tolerance of rapid freezing. Although freezing by itself is no barrier to survival, the metabolism and therefore reproduction of all life has an absolute requirement for liquid water. The rate and manner at which living organisms are cooled and thawed, however, are crucial factors in their ultimate survival frequencies.

R. Morita was concerned with field studies of organisms, especially microbes, found in extreme environments. How little definitive work there has been, for example, on both low and high temperature habitats, especially when they involve several environmental constraints at once, e.g. heat and high pressure, or high salt. Although some progress has been made in the study of life in hot springs, most of the organisms have not been isolated, characterized or thus identified, nor have the mechanisms of tolerance been analyzed. Such work is now beginning. In one study, by isolating, purifying and measuring the properties of an enzyme, malate dehydrogenase from a bacillus found in very hot environments, H. Kristjansson and C. Ponnamperuma showed that heat-induced conformational changes in the enzyme - 51 and 59 °C - may be related to mechanisms of growth at very hot temperatures.

Not only are microbes found in many environments with extremely high or low concentration of mineral ions in them, but they also may interact strongly with such inorganic ions, thus altering their immediate surroundings. K. H. Nealson presented evidence where thick coverings of manganese dioxide surround both the vegatative cells and spores of bacteria capable of the oxidation of soluble manganese into an MnO_2 precipitate. He pointed out that the continued recognition of the role of microbes in catalysing the transformation of sediments and aqueous environments in which they are found is a prerequisite for the interpretation of paleoenvironments - especially those that prevailed during the accumulation of large quantities of metalliferous sediments.

One of the ways in which microbes have adapted to extreme conditions is by moving into less demanding habitats. J. Staley argues that the gas vacuole, a bacterial organelle that regulates vertical position of the organism in the water column by altering cell buoyancy, evolved very early. Both the gas vacuole and the flagella may represent the earliest behavioral mechanisms to have evolved for avoidance of rather than the tolerance of environmental insults.

These published papers were supported by a number of shorter contributed papers.

A Colloquium of this size cannot be held without the assistance of a great number of people whose names never appear in print. The staff of the Laboratory of Chemical Evolution, particularly Mrs. Detra Rose, put in uncounted hours overseeing and taking care of the little details that made this program run smoothly. In the publication of these Proceedings, acknowledgement must be made of the contributions of Johnet J. Kemper, Faculty Research Assistant in the Laboratory of Chemical Evolution. With efficiency, skill and an incomparable sense of humor and perspective, Johnet relieved the editors of the task of managing these pages from first submission through final publication.

EDITORS

NOTES

[1] C. Ponnamperuma (ed.), *Chemical Evolution of the Giant Planets,* Academic Press, New York, 1976.
[2] C. Ponnamperuma (ed.), *Chemical Evolution of the Precambrian,* Academic Press, New York, 1976.
[3] C. Ponnamperuma (ed.), *Comparative Planetology,* Academic Press, New York, 1978.

LIMITS TO LIFE AT LOW TEMPERATURES AND AT REDUCED WATER CONTENTS AND WATER ACTIVITIES

PETER MAZUR

Biology Division, Oak Ridge National Laboratory,* Oak Ridge, Tenn. 37830, U.S.A.

Abstract. Liquid water is generally considered an absolute requisite for functional life; consequently, life is expected to function only over the range of temperatures that permit its existence. These limits, however, do not apply to cell survival. Some cells can survive the closest attainable approach to 0 K, and some can survive the loss of over 99% of their water.

1. Cell Survival at Subzero Temperatures

Many cells that can survive the initial trauma of freezing die with time at temperatures above about −80 °C. The rate of death is dependent in a complex way on temperature, on the species and type of cell, and on the composition of the medium in which they are frozen. The killing rate can vary from several percent per hour to several percent per year (Mazur, 1966). Below about −130 °C, on the other hand, no thermally driven reactions can occur in aqueous systems, and there are no confirmed instances of progressive loss in viability (Mazur, 1976).

Liquid water does not exist below −130 °C; the only states of water that do exist are crystalline or glassy, and in both cases the viscosity is so high ($>10^{13}$ poises) that diffusion is not significant over time spans of human interest. In addition to this diffusional barrier, there is insufficient thermal energy for most reactions. A few nonaqueous reactions may occur at −196 °C, but only between highly reactive compounds of O, H, and F in the presence of an exogenous source of energy such as a glow discharge (McGee and Martin, 1962).

The only reactions that can occur in frozen aqueous systems at −196 °C are photophysical events such as the formation of free radicals and the production of breaks in macromolecules as a direct result of 'hits' by background ionizing radiation or high-energy particles (Rice, 1960). It is conceivable that, over a sufficiently long period of time, these direct ionizations could produce enough breaks or other damage in DNA to become deleterious after rewarming to physiological temperatures, especially since no enzymatic repair can occur at these very low temperatures. One can estimate the death rate at −196 °C from radiation data at physiological temperatures on cells irradiated under anoxic conditions at high dose rates (∼100 rad min^{-1}). (At high dose rates, the irradiation is probably completed before significant repair can be initiated, and in that sense it is

---

* Operated by Union Carbide Corporation under contract W-7405-eng-26 with the U.S. Department of Energy.

analogous to the long-duration, low-rate irradiation that cells are subjected to at −196 °C. Irradiation under anoxic conditions reduces the biochemical component of ionizing radiation (Wood and Taylor, 1957).) The dose of ionizing radiation that kills 63% of representative cultured mammalian cells is 200 to 400 rad (Elkind and Whitmore, 1967). Since terrestrial background radiation is some 0.1 rad yr^{-1}, it ought to require some 2000 to 4000 yr at −196 °C to kill 63% of a population of typical mammalian cells. The time period would be about 200 yr for the most sensitive known mammalian cells, the early mouse oocyte, and it would be well over a million years for highly radioresistant organisms like *Micrococcus radiodurans*. Analogous arguments apply to mutation rates at −196 °C. They ought to be greatly reduced (see Mazur, 1976, for references).

The fundamental challenge to cell survival, then, is not low temperature *per se*; rather, it is whether cells can survive the initial cooling to the low temperatures and the subsequent return to physiological temperatures.

2. Lethal Events during the Freezing and Thawing of Cells

A. EFFECT OF COOLING RATE

When cells are cooled to −196 °C at various rates, the resulting survivals generally take the form shown in Figure 1. Maximum survival occurs at some intermediate optimal rate.

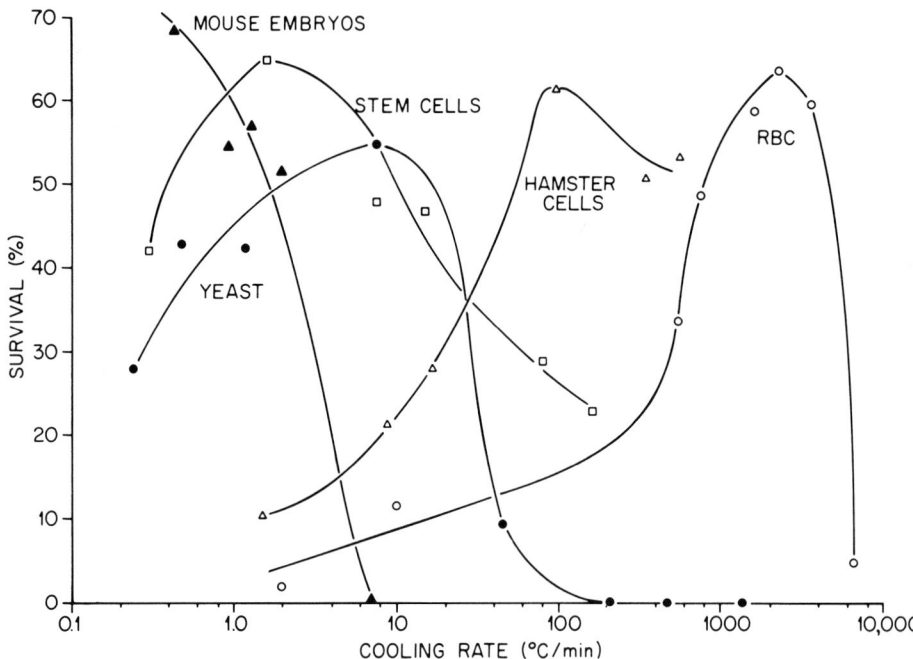

Fig. 1. Comparative effects of cooling velocity on the survival of various cells cooled to −196 °C. The data on mouse embryos, mouse marrow stem cells, yeast, hamster tissue culture cells, and human erythrocytes are from Whittingham *et al.* (1972), Leibo *et al.* (1970), Mazur and Schmidt (1968), Mazur *et al.* (1969), and Rapatz *et al.* (1968).

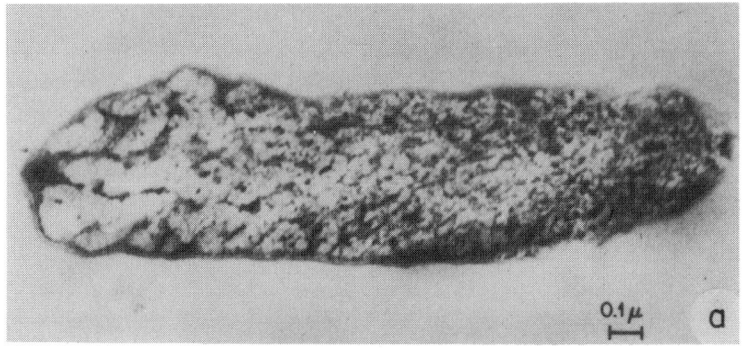

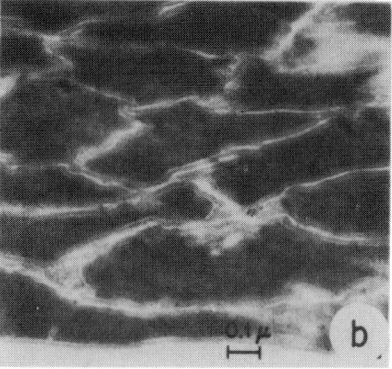

Fig. 2. Microscopic appearance of *Escherichia coli* after rapid freezing (a) or slow freezing at 1 °C min (b). The cells were freeze-substituted at −78 °C with ethanol. (Modified from Rapatz *et al.*, 1966).

Cooling at lower rates is usually significantly more deleterious, and cooling at higher rates is usually dramatically more injurious. However, the numerical value of the cooling rate that produces maximum survival can vary widely among cell types. In the examples shown, it ranges from <1 °C min^{-1} for mouse ova to about 1000 °C min^{-1} for human red cells.

Some clues as to the determinants of the inverted-U curves can be obtained from an examination of the morphological appearance of cells frozen at various rates. Figure 2 shows freeze-substituted *Escherichia coli*. Rapidly cooled cells (Figure 2a) are relatively normal in shape and size, but they are pitted with intracellular spaces that represent the former location of ice crystals. Slowly frozen *E. coli* (Figure 2b), on the other hand, are highly shrunken and show no evidence of intracellular ice.

Slow and rapid cooling produce analogous effects on mouse ova (Figure 3). As noted in Figure 1, the optimum cooling rate for ova is ≤1 °C min^{-1}. From Figure 3 we see that those frozen rapidly at 32 °C min^{-1} undergo little shrinkage and they freeze intracellularly at about −40 °C, whereas those frozen slowly at close to the optimum rate (1.2 °C min^{-1}) become progressively shrunken and show no evidence of intracellular ice.

A close correlation has been found in three types of cells between the cooling rates

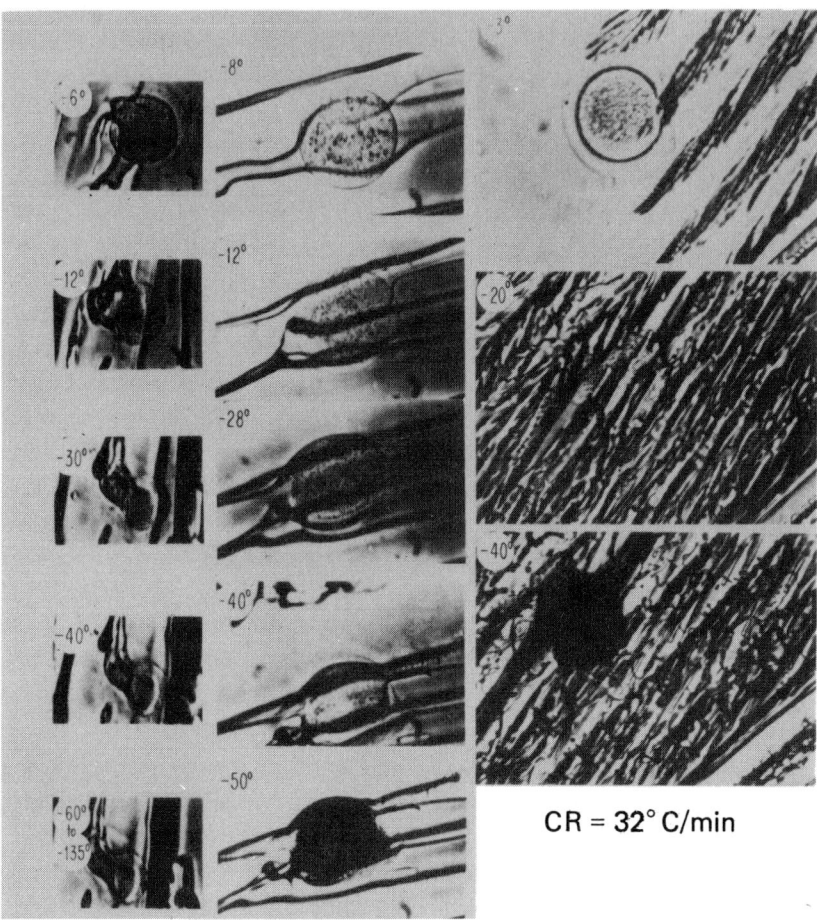

Fig. 3. Effects of cooling rate on the morphology of unfertilized mouse ova during freezing in $1M$ dimethyl sulfoxide. The cells cooled at $1.2\,°C\,min^{-1}$ dehydrate without intracellular freezing. Those cooled at 2.4 or $32°C\,min^{-1}$ undergo little shrinkage during cooling and eventually freeze intracellularly between $-40°$ and $-50°C$. Intracellular freezing is manifested by the sudden 'blacking out' of the cell as a result of light scattered by numerous small crystals. (Modified from Leibo et al., 1978.)

that are supraoptimal in terms of survival and the cooling rates that produce intracellular ice (Figure 4).

The essential events that occur during freezing are shown in schematic form in Figure 5. Most cells remain unfrozen at -10 to $-15\,°C$ even though those temperatures are 9 to 14 degrees below the actual freezing point of their cytoplasm, and even though ice is present in the outside medium. Apparently the cell membrane is able to block the passage of extracellular ice above about $-15\,°C$, and thus is able to prevent the nucleation of the supercooled water within the cell by that ice. Explanations for the ability of membranes to block nucleation have been suggested (Mazur, 1977a).

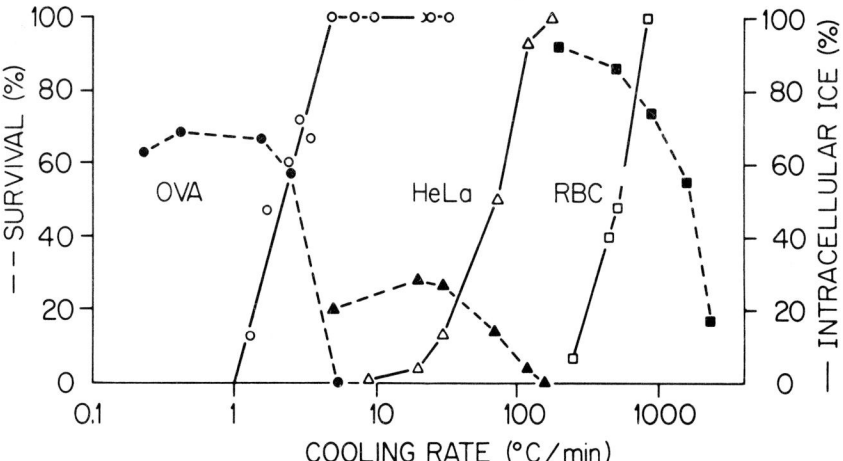

Fig. 4. Percentage survival (– – –) plotted against the percentage of cells undergoing intracellular freezing (——) in three mammalian cells frozen at various rates to –20 °C (HeLa) or –196 °C. (Modified from Leibo, 1977.)

Supercooled water by definition has a higher vapor pressure (activity or chemical potential)* than that of ice outside. Since water will move from regions of high activity to regions of low activity, and since the membrane is permeable to liquid water (although not ice crystals), water will move out of the cell and freeze externally. This efflux concentrates the intracellular solution and lowers its activity (a_w).

The end result depends on the cooling rate. If the cooling rate is sufficiently slow, the cell will be able by progressive dehydration to maintain the intracellular a_w close to the a_w of the external ice and external solution. Under these conditions intracellular freezing is unlikely. But if the cooling is not sufficiently slow, the cell will not be able to lose water fast enough to reduce the intracellular a_w to the equilibrium value. If the cell remains supercooled below –15 °C, it will eventually freeze intracellularly.

The term 'sufficiently slow' can be quantified. Were no water movement to occur, the vapor pressure of ice outside the cell (p_e) would decrease faster than that of the supercooled water inside the cell (p_i), the two vapor pressures being given by

$$\frac{d \ln p_e}{dT} = \frac{L_s}{RT^2}, \tag{1}$$

$$\frac{d \ln p_i}{dT} = \frac{L_1}{RT^2} + \frac{d \ln x_i}{dT}, \tag{2}$$

respectively. (L_s and L_v are the latent heat of sublimation and vaporization, x_i is the mole fraction of intracellular water, R is the gas constant, and T is temperature.) But the very

---

*Water activity is defined as $a_w = p_{H_2O}$ (solution)/p_{H_2O} (liquid pure) = p_{ice}/p_{H_2O} (liquid pure), where p is vapor pressure. Chemical potential (μ_w) is $\mu_w = \mu_w^\circ + RT \ln a_w$, where μ_w° is the chemical potential of pure water.

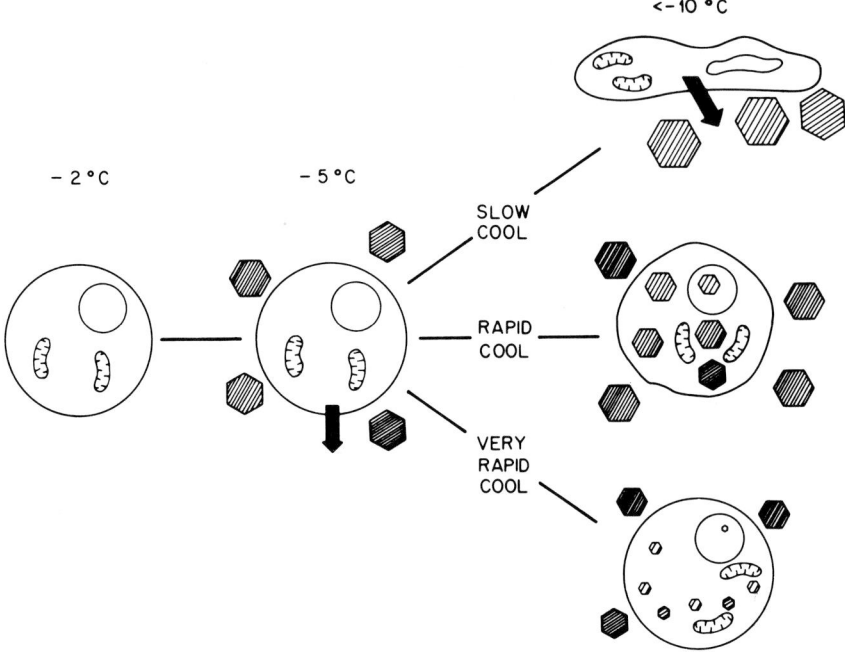

Fig. 5. Schematic of physical events in cells during freezing. The cross-hatched hexagons represent ice crystals. (From Mazur, 1977a.)

development of a difference in vapor pressure produces a proportional force that drives water out of the cell at a rate

$$\frac{dV}{dt} = \frac{KART}{v_1^\circ} \ln p_e/p_i, \qquad (3)$$

where V is the volume of cell water, K is the permeability of the cell to water (hydraulic conductivity), and A its surface area. Equations (1), (2), and (3) along with an exponential equation describing the temperature dependence of K, and an equation relating time, t, and temperature, T (i.e., the cooling rate), can be solved numerically to give the volume of cell water remaining in the cell as a function of temperature for given values of cooling rate, permeability, and cell surface area (Mazur, 1963a, 1977a). The equations have been solved for various cells. Calculated solutions for yeast cooled at various rates are shown in Figure 6. Curve 'Eq' expresses the reduction in the relative volume of intracellular water (relative to the normal volume) that is required to maintain equilibrium between the activities of cell water and outside ice. Departures from the equilibrium curve mean by definition that the cell water is supercooled. The calculations indicate that the water contents of yeast cooled at 1 or 10 °C min^{-1} will reach the equilibrium value at $-7°$ or above. We would predict therefore that cells cooled at 1 or 10 °C min^{-1} will not freeze intracellularly. On the other hand, yeast cells cooled at 100 °C min^{-1} or faster are far from equilibrium at -15 °C, the temperature at which most supercooled cells nucleate.

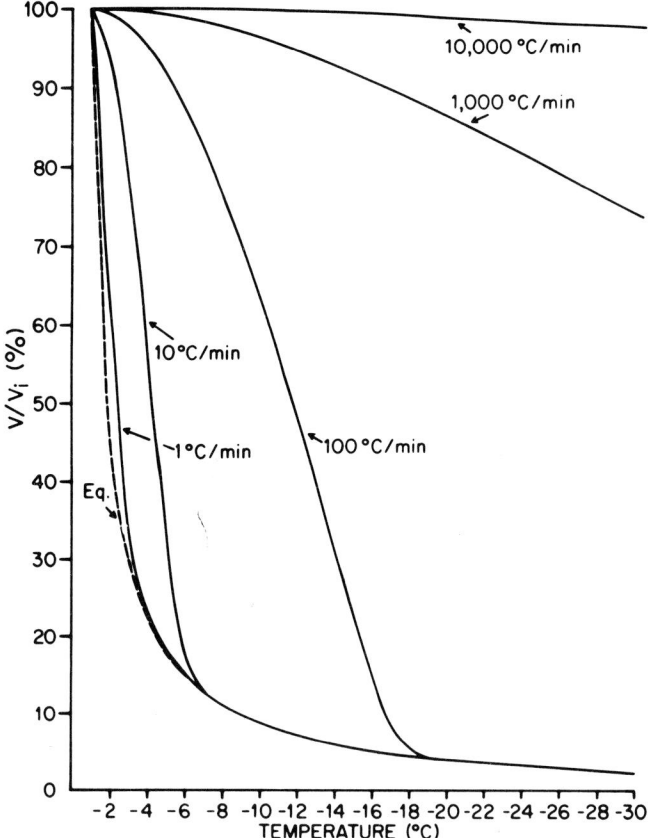

Fig. 6. Calculated percentage of intracellular water remaining in yeast at various subzero temperatures after cooling at the indicated rates. Eq. is the calculated equilibrium curve for cells cooled at infinitesimal rates. (From Mazur, 1963a.)

We would predict, therefore, that yeast cooled at 100 °C min^{-1} or faster will freeze intracellularly. Figure 7 expresses these predictions as a plot of the probability of intracellular freezing versus cooling rate, and it compares the probabilities with the observed survivals. Clearly the cooling rates predicted to produce intracellular ice coincide closely with the cooling rates that cause the abrupt drop on the right-hand side of the survival curve. Considerations of this sort, along with direct microscopic observations on yeast and other cells relating frequency of intracellular freezing to cooling rate (e.g., Figure 4), have provided strong support for the following conclusion: The drop in survival at high cooling rates is a consequence of the formation of intracellular ice (and, as we shall see shortly, of events during slow warming). The numerical value of the critical cooling rate can vary widely (see Figure 1). This variation is chiefly a consequence of differences in the surface area, and in the permeability coefficient for water and its temperature coefficient.

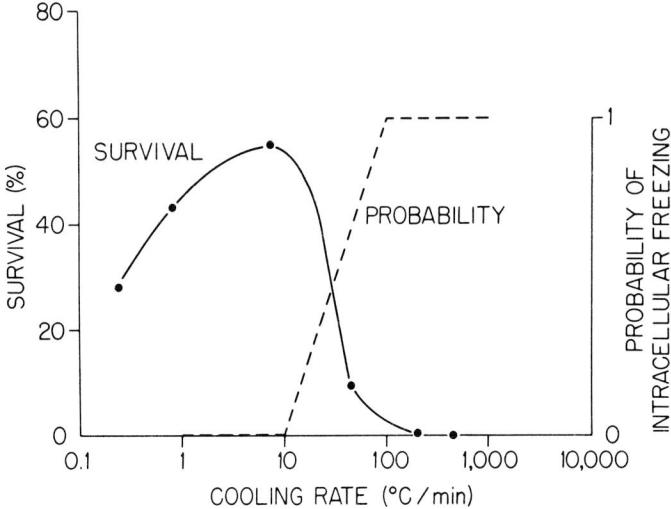

Fig. 7. Comparative effects of cooling rate on the survival and probability of intracellular freezing of yeast. (From Mazur, 1977a.)

B. CONSEQUENCES OF SLOW FREEZING

The data in Figure 7 as well as those in Figure 1 also illustrate that too low a cooling rate is injurious. To what can this slow freezing injury be ascribed? Since slow freezing injury is associated with modifications in the composition and properties of aqueous solutions produced by extracellular ice, it is sometimes referred to as 'solution-effect' injury. On a molar basis, the principal solutes in and around cells are electrolytes, and one consequence of progressive freezing is a dramatic increase in the solute concentration in the residual unfrozen puddles. For example, as shown in the left- hand curve of Figure 8, when a $0.15M$ solution of NaCl is frozen to $-10\,°C$, the concentration of NaCl in the unfrozen portion of the solution rises some 20-fold to 2.8 molal. (It is this high solute concentration that causes the drop in extracellular a_w discussed in the previous section.) Mammalian cells suspended in that solution will generally be killed by about $-20\,°C$ (Figure 9). This correlation between cell death and high salt concentration led Lovelock (1953) and succeeding investigators to ascribe a causal relationship between the two.

Three years previously, Smith and Polge (1950) had discovered that the addition of about $1M$ glycerol to a suspension of spermatozoa in saline resulted in a high proportion of the cells surviving slow freezing to $-80\,°C$. The marked protective effect of glycerol was subsequently found for many cells, and is illustrated for mouse marrow cells in Figure 10. Inspection of Figure 8 shows that the ability of glycerol to protect viability is paralled by its ability to suppress the concentration of salts at a given temperature. For example, when a solution of 0.15 molal salt in $1.0M$ glycerol is frozen to $-10\,°C$, the salt concentration rises only to 0.6 molal. (As noted, in the absence of glycerol it rises to 2.8 molal.) The suppression of salt concentration by glycerol is not a special property

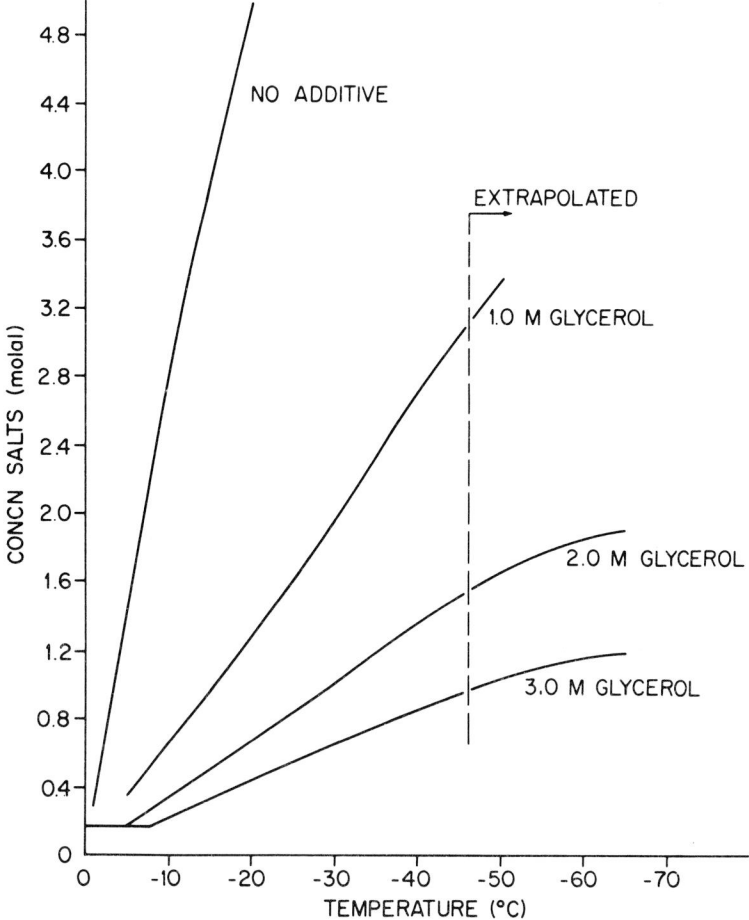

Fig. 8. Effect of the presence of glycerol on the concentration of salt in the unfrozen portions of a solution at various subzero temperatures. (Modified from Rall et al., 1978.)

of glycerol, but is rather a consequence of the phase rule and the colligative properties of solutes. The phase rule dictates that in a two-phase system (here, ice and liquid solution) at constant pressure the *total* solute concentration in the liquid solution is invariant at a given temperature. Since the total solute concentration is fixed, the higher the concentration of glycerol (or any other solute) the lower will be the salt concentration at that temperature. The effect is roughly proportional to the mole ratio of glycerol to salts in the medium. However freezing produces other changes in solutions besides an increase in solute concentration. In fact a current major area of cryobiological research is an attempt to define which changes are responsible for slow-freezing injury and which are prevented or counteracted by additives such as glycerol.

In the absence of protective compounds, mammalian cells are far more sensitive to slow freezing (solution effect injury) than are most microorganisms, although the latter

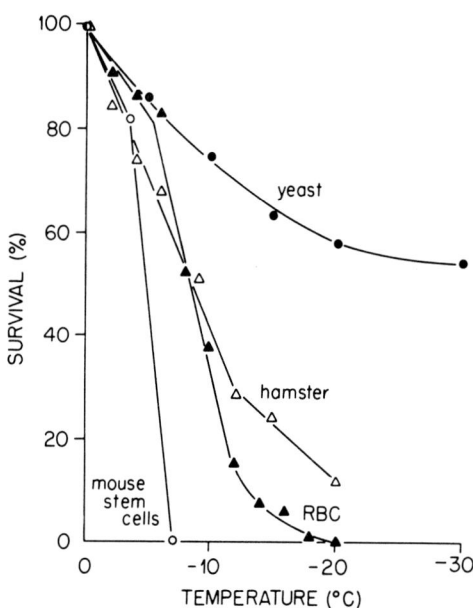

Fig. 9. Comparative sensitivity of four cells to slow freezing to various subzero temperatures in the absence of a protective compound. The data on yeast, hamster issue culture cells, human erythrocytes, and mouse marrow stem cells are from Mazur and Schmidt (1968), Mazur and Leibo (unpublished). Souzu and Mazur (1978), and Leibo et al. (1970), respectively.

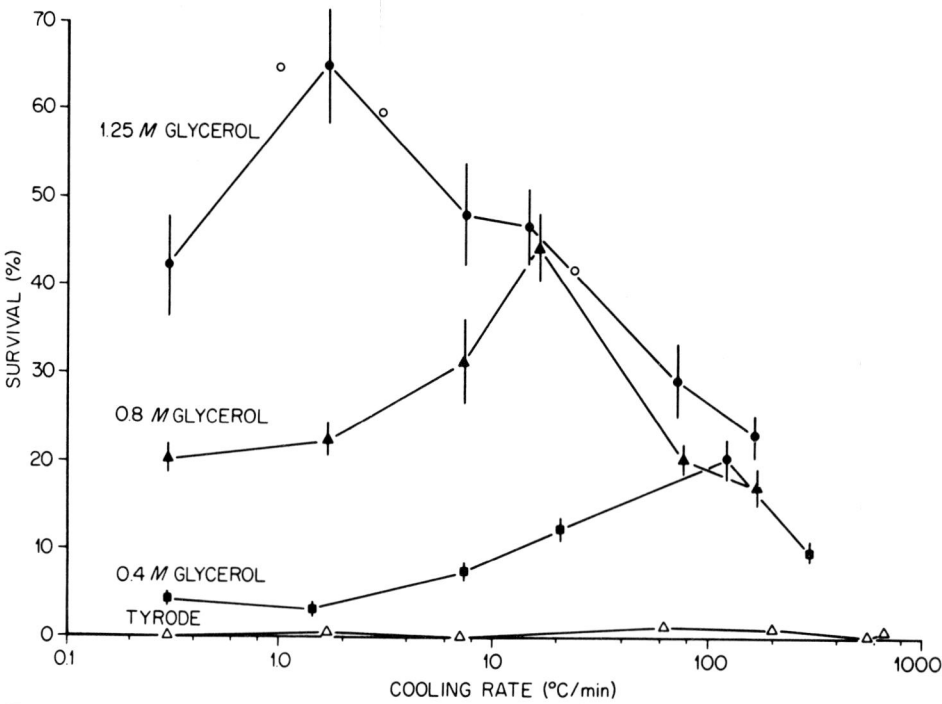

Fig. 10. Survival of mouse marrow stem cells suspended in balanced salt solution containing 0, 0.4, 0.8, or 1.25M glycerol after cooling to $-196\,°C$ at various rates and rapid warming. (Redrawn from Leibo et al., 1970.)

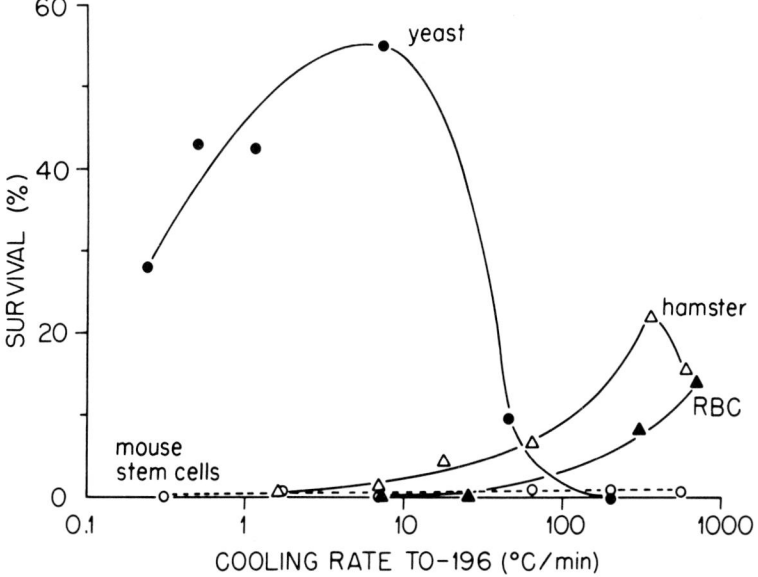

Fig. 11. Comparative sensitivity of four cells to freezing to −196 °C at various rates in the absence of a protective compound. The data on yeast, hamster tissue culture cells, human erythrocytes, and mouse marrow stem cells are from Mazur and Schmidt (1968), Mazur et al. (1969), Morris and Farrant (1972), and Leibo et al. (1970), respectively.

are by no means immune (Figures 9 and 11). Both mammalian cells and microorganisms are, however, susceptible to the intracellular freezing that results from cooling at supraoptimal rates.

C. EFFECT OF WARMING RATE ON SURVIVAL

The effects of rate of thawing on cell survival are strongly influenced by the prior rate of cooling. In almost all cases examined, cells that are cooled at supraoptimal rates (i.e., rates sufficient to induce intracellular ice) survive better when warming is rapid than when it is slow. The dramatic effect of warming rate on yeast cells is illustrated in Figure 12. There is strong evidence, especially in yeast, that the high sensitivity of rapidly cooled cells to slow thawing results from the growth of intracellular ice crystals to damaging size by recrystallization during warming. This conclusion is based on the sort of evidence presented in Figure 13. In the experiment represented by the bottom graph, yeast were rapidly cooled to −196 °C, warmed slowly to the temperatures indicated on the abscissa, and then thawed rapidly. Slow warming up to −40 °C produced no damage, but continuing the slow warming to temperatures above −40 °C produced increasing injury. By −20 °C, 95% of the cells were killed. The yeast photographed in the upper half of the figure were subjected to somewhat analogous treatments. The cells were first cooled rapidly to −196 °C. They were then warmed abruptly to −50, −35, or −20 °C, held at these temperatures for various times, recooled to −196 °C, and freeze-cleaved below −100 °C. Increasing the intermediate holding temperature from −50 °C to −35 to −20 °C

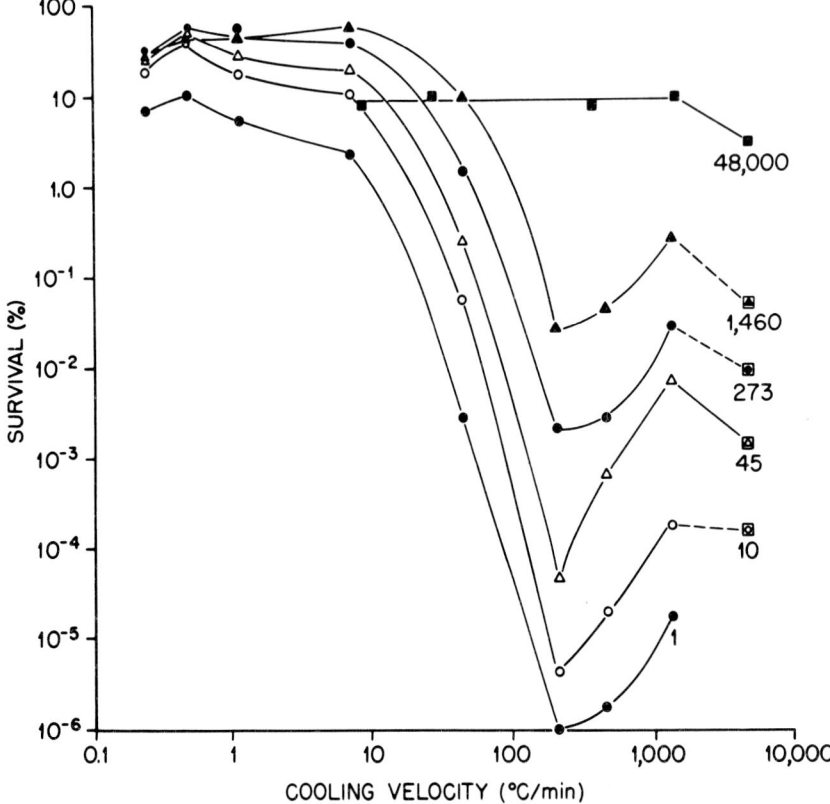

Fig. 12. Interactions of cooling rate and warming rate on the survival of frozen–thawed yeast. The numbers on the individual curves refer to warming rates. (From Mazur and Schmidt, 1968.)

produced progressively larger intracellular ice crystals. This increase in crystal size from recrystallization thus paralleled the decrease in survival shown in the lower half of the figure.

But what about the effect of rate of warming on cells cooled slowly enough to avoid intracellular freezing? In some cases there is little or no effect (e.g., yeast cells, Figure 12). But in other cases there are major effects quite different from those observed in rapidly cooled cells. In human erythrocytes, for example (Figure 14), the sequence of slow freezing (0.5 °C min^{-1}) followed by slow thawing (0.5 to 1 °C min^{-1}) is considerably less damaging than the sequence of slow freezing followed by rapid thawing (160 °C min^{-1}).

Although it is not known whether sensitivity of slowly cooled cells to rapid thawing is a characteristic of all cells, it is characteristic of groups as disparate as erythrocytes, mouse embryos (Leibo et al., 1974), and plant cells (Levitt, 1966). Although its basis is not understood, the response is similar to that seen in osmotic shock. During the progressive thawing that accompanies warming, the highly concentrated extracellular medium is progressively diluted by melting ice. Hence, sensitivity to rapid thawing may be tantamount to sensitivity to rapid dilution, and the latter by definition is osmotic shock. I have

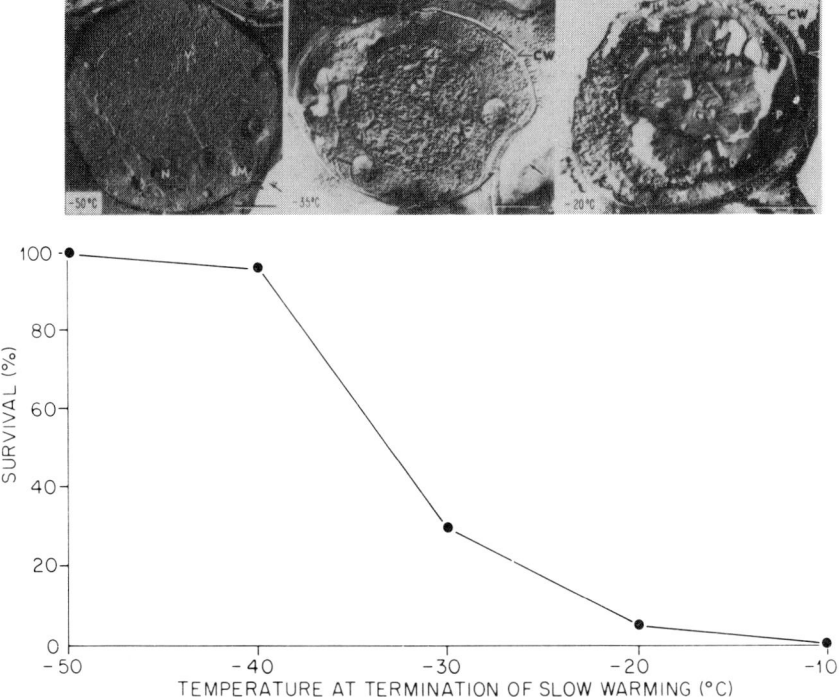

Fig. 13. Effect of warming to various temperatures on survival and on the recrystallization of intracellular ice in yeast rapidly frozen to −196 °C. The data are from MacKenzie (1970) and Bank (1973). In MacKenzie's survival studies the cells were warmed slowly to the indicated temperatures and then thawed rapidly. In Bank's recrystallization studies, the cells were warmed abruptly to −50, −35, and −20°C; held 24 hr, 30 min, and 5 min, respectively; recooled to <−100°C; and freeze-cleaved.

speculated that shock may occur when excessive additive driven into cells during cooling cannot diffuse out fast enough during warming to prevent osmotic water influx and consequent cell swelling (Mazur, 1977b).

3. Consequences of Cell Dehydration

Many cells in nature undergo periodic or intermittent dehydration, and in the laboratory dehydration is one important method for preserving cultures (Heckly, 1961). The loss of cell water has a number of interrelated but distinguishable physical-chemical consequences. Since water is the only volatile compound present in cells in more than trace amounts, one effect of dehydration is to increase the concentration of intracellular solutes thereby lowering the water activity, a_w. A second effect is to alter the composition of the cytoplasm as various solutes begin to precipitate. Differential precipitation can produce rather major changes in pH. Finally, as more and more water is removed, the structure and properties of the remaining liquid begin to differ sharply from that of normal water.

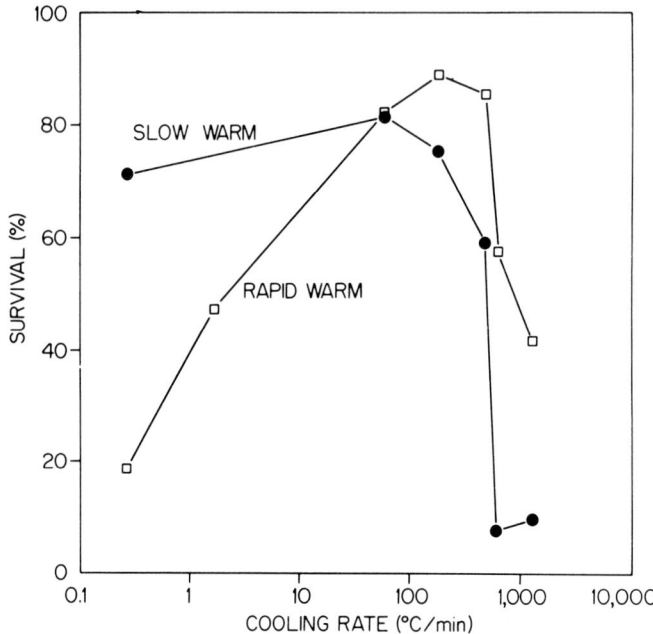

Fig. 14. Survival (percentage unhemolyzed) of human erythrocytes suspended in 2M glycerol, frozen at various rates to −196 °C, and warmed slowly (0.5 °C min^{-1}) or rapidly (160 °C min^{-1}). (Modified from Miller and Mazur, 1976.)

As mentioned, these phenomena are to a degree separable. A halophilic bacterium in 4M NaCl, for example, is not dehydrated, but it is exposed to low water activity (a_w = 0.84). On the other hand, a cell drifting in the atmosphere on a humid day will be extensively dehydrated even though the relative humidity ($\equiv 100 \times a_w$) may be 97%.

A. FREEZING VERSUS DEHYDRATION AND BOUND WATER

The dehydration produced by slow freezing is self-evident (see Figures 2 and 3). But rapid freezing also produces the equivalent of cell dehydration. The difference is that in rapid freezing liquid cell water is sequestered as ice in the cell's interior, whereas in slow freezing it is sequestered as ice in the external medium.

About 90% of the water of yeast and *E. coli* is frozen by −20 °C. The remaining 10% remains unfrozen no matter how low the temperature (Figure 15). It is by definition 'bound'. These conclusions are based on calorimetric measurements by Wood and Rosenberg (1957), Souzu *et al.* (1961), and Mazur (1963b). A more precise wording would be that 10% of the cell water does not exhibit a latent heat of fusion. Expressed in grams of water per gram of cell solids, the unfreezable fraction is about 0.25 g g^{-1} or about 0.20 g g^{-1} wet weight.

The inability to freeze is not the only unusual characteristic of this bound water fraction. Koga *et al.* (1966) have shown that when the water content of yeast is reduced below these figures, the dielectric constant and NMR signal intensities drop sharply

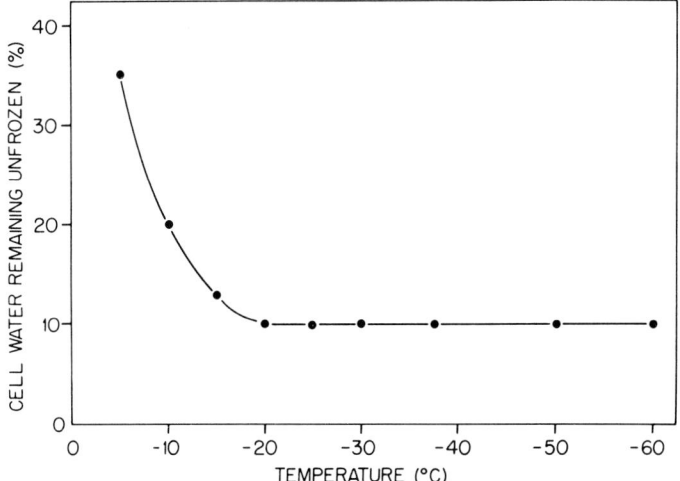

Fig. 15. Calorimetric estimates of the percentage of yeast cell water frozen at various temperatures. (From data of Wood and Rosenberg, 1957.)

(Figure 16). By the time the water content is reduced to about 0.1 g H_2O g^{-1} wet weight, it appears to be irrotationally bound.

Although the bound fraction cannot be converted to ice, it can be removed from cells by exposing them for a sufficiently long time to a sufficiently low a_w. The water content of a system after equilibration with various external a_w is a sorption isotherm, and Koga et al. (1966) have also obtained the sorption isotherm of yeast (Figure 17). The sigmoid-shaped curve is typical for cells and for cell constituents like proteins and nucleic acids.

B. EFFECTS OF DEHYDRATION ON CELL SURVIVAL

Much information on the effects of dehydration on cells comes from laboratory studies on freeze-drying. In this procedure, cell suspensions are frozen and placed under high vacuum in communication with a vapor trap. The trap, usually a condenser held at −196 °C, maintains the a_w of trapped water near zero. Water sublimes from the sample and condenses in the trap, so that after some hours the water content of the sample will be reduced to ≤1 to 2% of the original. The high vacuum accelerates the sublimation by increasing the mean free path of the water vapor molecules.

Freeze-drying is almost invariably more damaging than freeze-thawing, a point illustrated in Table I for a number of fungi. It is even harsher to mammalian cells. There is no confirmed instance of any surviving freeze-drying to low water contents. The greater lethality of freeze-drying appears to result from the removal of a portion of the cell's bound water. Sakurada (1958), for example, found that 50% of a population of yeast withstood having their water content lowered to 20% of normal, but only 5% survived the removal of the remaining 20%. Nei et al. (1965) report similar findings for E. coli. Interestingly, as increasing fractions of the bound water of yeast are removed, the survival of the cells depends increasingly on the temperature of the rehydrating medium. For

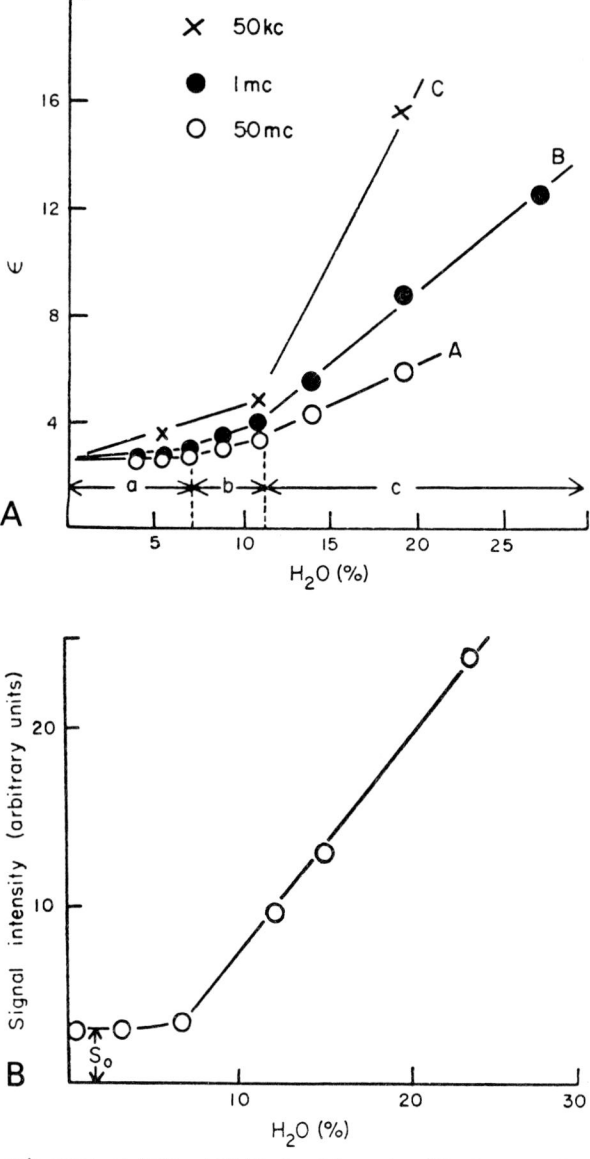

Fig. 16. Dielectric constant (A) and NMR signal intensity (B) of yeast cells as a function of their water content ($100 \times $ g H_2O g^{-1} wet weight). (From Koga et al., 1966.) (The dielectric constant of pure bulk water and ice at ≤ 50 MHz are about 81 and 3, respectively.)

unknown reasons, rehydration by water at 4 °C kills as many as 95% of the cells that have had some 96% of their water removed (Sant and Peterson, 1958), whereas rehydration at ≥ 26 °C does not.

Another approach to examining the effects of dehydration on cells is to equilibrate samples in atmospheres at various a_w's and to determine the subsequent rate of killing.

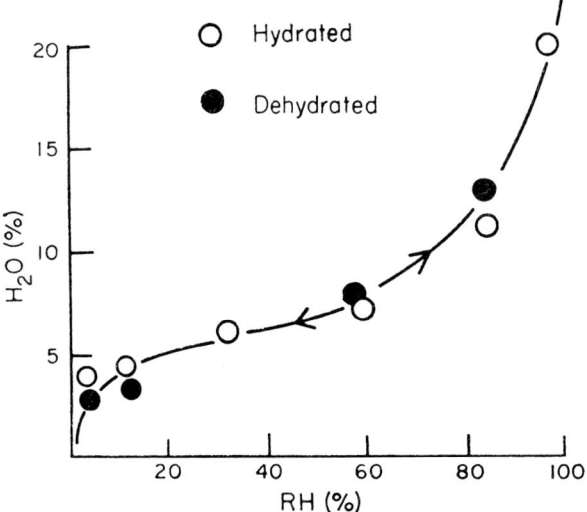

Fig. 17. Sorption isotherms of yeast cells; i.e., the water content of the cells (g H_2O g^{-1} wet weight) versus the relative humidity with which they were equilibrated. RH (%) ≡ 100 (a_w). (Data from Koga et al., 1966.)

Figure 18 summarizes data on the bacterium *Serratia marcescens*. As shown in the right-hand curve, a reduction in water activity from 1.0 to 0.9 reduces the cell water content to 10% of normal. But the consequent rate of killing (left curve) is low, a finding consistent with those for yeast (above), and consistent with the view that little bound water has been removed. However, reducing the a_w from 0.9 to 0.2 drops the cell water content from 10 to 1% and increases the rate of killing by a factor of 10. The general pattern is not unique to Serratia. Other microorganisms also exhibit increased death rates at humidities below 90% (Webb, 1967a; Chapman et al., 1967; Mackenzie, 1971; Cox, 1971;

TABLE I
Comparative effects of freeze-thawing and freeze-drying on the survival of Fungi[a]

| | | | % survival[b] after | |
Organism	Cell type	Freezing temperature (°C)	Freezing and thawing	Freeze-drying
Aspergillus flavus	Conidia	−15	65	26
		−65	7	2
A. niger	Conidia	−65	10	9
Fusarium	Conidia	−65	36	0
Gliocladium	Conidia	−65	73	0
Penicillium	Conidia	−65	57	2
Pestalotia	Conidia	−65	27	6
Saccharomyces cerevisia	Veg. Cells	−12	60	1

[a] Data from various authors (see Mazur, 1968).
[b] Survivals of the spores refer to % germination.

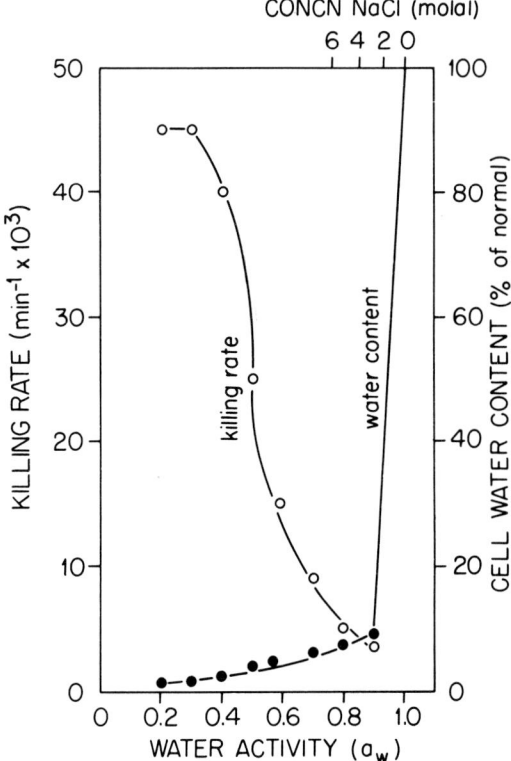

Fig. 18. Effect of water content and water activity on the rate of killing of cells of the bacterium *Serratia marcescens*. Note the concentrations of solute (NaCl) required to achieve various water activities. (From data of Webb, 1960; Bateman et al., 1962.)

Ehresmann and Hatch, 1975). In addition, death rates at lower water activities are influenced rather strongly by the partial pressure of oxygen in the atmosphere. Higher oxygen pressures are usually more damaging (Zentner, 1966; Webb, 1967a). Why the removal of bound water is damaging to cells is by no means clear. The bound water fraction of cells (i.e., ~0.3 g H_2O g^{-1} solids) is comparable to the amount of water in the hydration layer of biological macromolecules (Kuntz et al., 1969); so perhaps the question can be rephrased to ask why is the removal of water of hydration damaging to cells? There is speculation but little in the way of answers.

4. Limits to Survival at Low Temperatures and to Dehydration

Any cell that tolerates freezing to about −70 °C appears capable of withstanding the lowest temperatures attainable. Even complex forms like early mouse embryos survive freezing to −269 °C without difficulty (Whittingham et al., 1972). Nor does there appear to be a fundamental limit to the degree of dehydration that is tolerable by at least some cells in the absence of oxygen. Bacterial forms have been dried to water activities that approach zero (Portner *et al.*, 1961; Morelli *et al.*, 1962). Still, although there are no absolute limits, organisms exhibit an enormous range in their sensitivity to low temperatures and dehydration, and the sensitivity of a given organism can vary by orders of magnitude depending on how it has been frozen and thawed or how it has been dehydrated.

Conditions in nature reflect the close coupling between freezing and dehydration in that forms that tolerate the latter will generally tolerate the former (Levitt, 1966). Examples are fungous conidia, bacterial spores, and Artemia cysts. But the reverse is not necessarily true. Numerous forms that in nature or the laboratory are highly sensitive to dehydration will survive freezing under appropriate conditions. Examples include many protozoa and apparently all mammalian cells. Finally, there are forms in nature (higher plants and insects) that undergo remarkable seasonal increases in their ability to tolerate freezing (Levitt, 1966; Li and Sakai, 1978).

The fraction of cells that survives freezing is generally no more resistant to subsequent freezing and thawing than the original population. This fact plus other evidence indicates the survivors of freezing are not a genetic subset of the original population. In other words, freezing is neither selective nor mutagenic. On the other hand, there is evidence that freeze- and air-drying can be mutagenic (Webb, 1967; Servin-Massieu and Cruz-Camarillo, 1969; Hieda and Ito, 1973; Ashwood-Smith and Grant, 1976). However, its mutagenicity is apparently not severe enough to cause obvious problems for those who use the procedure to preserve cultures of microorganisms (Brown, 1963).

5. Minimum Temperatures for Cell Growth and Function

A. EFFECT OF SUBZERO TEMPERATURES *PER SE*

Some information on the effects on cells of subzero temperature *per se* (i.e., in the absence of freezing) can be obtained by taking advantage of the ability of dilute aqueous solutions to supercool to about $-10°$ and occasionally to as low as $-20\,°C$. Most (although not all) cells survive supercooling to such temperatures, but relatively little is known about their ability to grow and function for extended periods of time. One fascinating exception is the report of Scholander *et al.* (1957) and Scholander and Maggert (1971) who discuss certain arctic fish that live permanently supercooled a degree or so. Presumably, cell responses a few degrees below $0\,°C$ ought not to be qualitatively different from their responses at $0\,°C$ or a few degrees above, and the responses to the latter been reviewed extensively (Baross and Morita, 1978; Inniss and Ingraham, 1978).

B. CELL GROWTH IN THE PARTLY FROZEN STATE

Unlike cell survival, there is a definite lower temperature limit for cell growth in the frozen state. Michener and Elliott (1964) thoroughly reviewed the situation for microorganisms and found that the number of reported instances of cell growth diminished rapidly a few degrees below $0\,°C$ (Figure 19). In fact, they report no confirmed cases of growth below $-12\,°C$. The inability of organisms to grow below $-12\,°C$ is consistent with the known physical state of aqueous solutions at these temperatures. As Table II shows, when

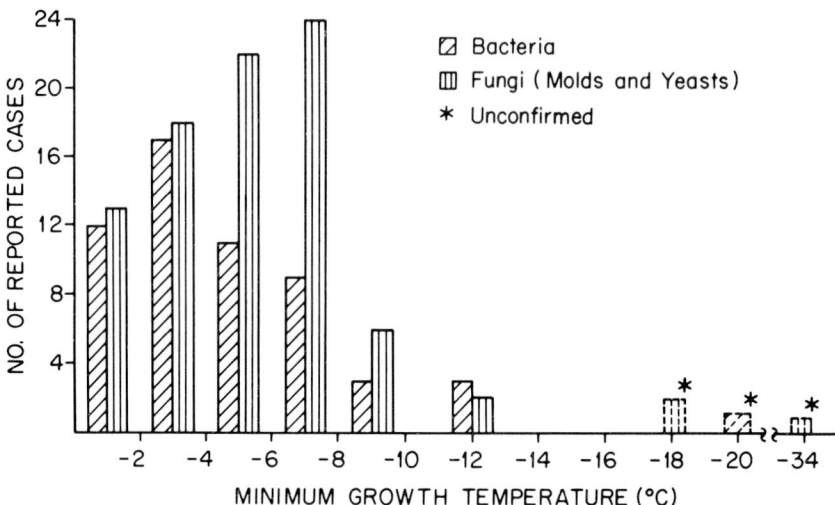

Fig. 19. Reported cases of microbial growth below 0 °C. Adapted from Michener and Elliott (1964) by Mazur et al. (1978).

solutions of sodium chloride in water, for example, are equilibrated at various subzero temperatures, the concentrations in the unfrozen portions reach 3.7 molal (3.5 molar) by −14 °C. For solutes in general, the concentrations of solutes in the unfrozen portions of solutions are given by $\phi \nu m = \Delta T/1.86$, where ϕ is the osmotic coefficient, ν the number of species into which the solute dissociates, and m is the molality. Aside from the toxic effects to nearly all microorganisms of such high concentrations of electrolytes, the high concentrations also depress the water activity (a_w) below the value permitting the growth even at optimal temperatures of nearly all microorganisms save halophilic and

TABLE II
Solute concentrations and water activities in NaCl solutions at various temperatures

Temperature (°C)	Concentration NaCl[a] (molal)	a_w [b]
−5	1.45	0.95
−10	2.79	0.91
−14	3.73	0.87
−16	4.17	0.85
−18	4.58	0.84
−20	4.99	0.82

[a] In the unfrozen portion of the solution. From International Critical tables (1926).

[b] Calculated from data of Dorsey (1940). The activities depend only on temperature (and total hydrostatic pressure); they are independent of the nature of the solute.

osmophilic forms (see below). As shown in Table II, the value of a_w at $-16\,°C$ has dropped to 0.85.

7. Cell Growth at Low Water Activities

The minimum water activities (a_w) for cell growth at normal temperatures have recently been reviewed by Rose (1976). Table III is a portion of his data. Most bacteria require an a_w above 0.9; most fungi an a_w above 0.85. The lowest limit is 0.61 for the fungus *Xeromyces bisporus*. The general lower limit of ~0.9 is not surprising. As just mentioned, a sodium chloride solution in equilibrium with an a_w of 0.9 has a concentration of about 3 molal (Table II).

I pointed out in the beginning of this chapter that cells can be suspended in *liquid* water at low a_w (lowered by high solute concentration), or they can be equilibrated with water *vapour* at any desired a_w between 0 and 0.99. There have been arguments over the years as to whether cell function and growth demand the presence of extracellular liquid water or whether high a_w water vapour would suffice. The preponderant opinion favors the former view. However, a careful study by Lange et al. (1970) indicates that lichens in the Negev Desert, which have become extremely dehydrated during the day,

TABLE III
Minimum water activities for growth of representative microorganisms
(From A. H. Rose, 1976)

Organism	Min a_w for growth	Equiv. conc. of NaCl at 25 °C[a] (molal)
Bacteria		
Aerobacter aerogenes	0.94	1.8
Bacillus cereus	0.99	0.3
Salmonella typhimurium	0.92	2.3
Micrococcus sp.	0.83	4.4
Halobacterium	0.76[b,c]	6.0
Yeasts		
Saccharomyces cerevisiae	0.92	2.3
Hansenula suaveolens	0.97	0.9
Saccharomyces rouxii	0.85[b]	4.0
Molds		
Rhizopus nigricans	0.93	2.0
Aspergillus niger	0.84	4.2
Aspergillus ruber	0.70	>6[d]
Xeromyces bisporus	0.61[b]	>6[e]

[a]From Robinson and Stokes (1959).
[b]Lowest value recorded for the indicated category of microorganism.
[c]The activity for a saturated solution of sodium chloride.
[d]Equivalent to 3.4 m $CaCl_2$.
[e]Equivalent to 4.0 m $CaCl_2$.

are capable at night of absorbing enough water vapor, even in the absence of the condensation of dew, to permit a short pulse of photosynthetic activity early in the morning. However the major portion of photosynthetic activity occurs when the lichen thalli are wet by condensed dew. Even in this remarkable case, then it appears that the organism may exhibit its full complement of functions only in the presence of liquid water.

References

Ashwood-Smith, M. J. and Grant, E.: 1976, *Cryobiol.* **13**, 206.
Bank, H.: 1973, *Cryobiol.* **10**, 157.
Baross, J. A. and Morita, R. Y.: 1978, in D. J. Kushner (ed.), *Microbial Life in Extreme Environments*, Academic Press, N.Y., pp. 9–71.
Bateman, J. B., Stevens, C. L., Mercer, W. B., and Carstensen, E. L.: 1962, *J. Gen. Microbiol.* **29**, 207.
Brown, W. E.: 1963, in S. M. Martin (ed.), *Culture Collections: Perspectives and Problems*, Univ. of Toronto Press, pp. 55–58.
Chapman, J. D., Webb, S. J., and Cormack, D. V.: 1967, *Nature* **213**, 465.
Cox, C. S.: 1971, *Appl. Microbiol.* **21**, 482.
Dorsey, N. E.: 1940, 'Properties of Ordinary Water Substance', ACS Monograph Series, No. 81, Reinhold Publ. Co., N.Y.
Ehresmann, D. W. and Hatch, M. T.: 1975, *Appl. Microbiol.* **29**, 352.
Elkind, M. M. and Whitmore, G. F.: 1967, *The Radiobiology of Cultured Mammalian Cells*, Gordon and Breach, Science Publ., N.Y.
Heckly, R. J.: 1961, in W. W. Umbreit (ed.), *Advances in Applied Microbiology*, Vol. 3, Academic Press, Inc., pp. 1–76.
Hieda, K. and Ito, T.: 1973, in *Freeze-drying of Biological Materials*, Proc. of C-1 Symp. (Sapporo), International Institute of Refrigeration, Paris, pp. 71–78.
Inniss, W. E. and Ingraham, J. L.: 1978, in D. J. Kushner (ed.), *Microbial Life in Extreme Environments*, Academic Press, Inc., N.Y., pp. 73–104.
International Critical Tables: 1926, McGraw-Hill, N.Y.
Koga, S., Echigo, A., and Nunomura, K.: 1966, *Biophys. J.* **6**, 665.
Kuntz, I. D., Jr., Brassfield, T. S., Law, G. D., and Purcell, G. V.: 1969, *Science*, **163**, 1329.
Lange, O. L., Schulze, E.-D., and Koch, W.: 1970, Experimentell-ökologische Untersuchungen an Flechten der Negev-Wüste. II. CO_2- Gaswechsel und Wasserhaushalt von *Ramalina Maciformis* (Del.) Bory am Natürlichen Standort während der Sommerlichen Trockenperiode. Flora. Bd. 159, pp. 38–62.
Leibo, S. P.: 1977, in D. Simatos, D. M. Strong, and J.-M. Turc (eds.), *Cryoimmunologie*, INSERM, Paris, pp. 311–334.
Leibo, S. P., Farrant, J., Mazur, P., Hanna, M. G., Jr., and Smith L. H.: 1970, *Cryobiol.* **6**, 315.
Leibo, S. P., Mazur, P., and Jackowski, S. C.: 1974, *Exp. Cell Res.* **89**, 79.
Leibo, S. P., McGrath, J. J., and Cravalho, E. G.: 1978, *Cryobiol.* **15**, 257.
Levitt, J.: 1966, in H. T. Meryman (ed.), *Cryobiology*, Academic Press, London, Chapter 11, pp. 495–563.
Li, P. H. and Sakai, A. (eds.): 1978, *Plant Cold Hardiness and Freezing Stress – Mechanisms and Crop Implications*, Academic Press, N.Y.
Lovelock, J. E.: 1953, *Biochim. Biophys. Acta.* **10**, 414.
MacKenzie, A. P.: 1970, in G. E. W. Wolstenholme and Maeve O'Connor (eds.), *The Frozen Cell*, Churchill, London, pp. 89–96.
Mackenzie, D. W.: 1971, *Appl. Microbiol.* **22**, 678.
McGee, H. A., Jr. and Martin, W. J.: 1962, *Cryogenics* **2**, 1.
Mazur, P.: 1963a, *J. Gen. Physiol.* **47**, 347.
Mazur, P.: 1963b, *Biophys. J.* **3**, 323.
Mazur, P.: 1966, in H. T. Meryman (ed.), *Cryobiology*, Academic Press, London, Chapter 6, pp. 213–315.

Mazur, P.: 1968, in G. C. Ainsworth and A. S. Sussman (eds.), *The Fungi*, Academic Press, N.Y., Vol. III, Chapter 14, pp. 325–394.
Mazur, P.: 1976, in Otto Mühlbock (ed.), *Proc. of Workshop on 'Basic Aspects of Freeze Preservation of Mouse Strains'*, Jackson Laboratory, Bar Harbor, Maine, Sept. 16–18, 1974, Gustav Fischer Verlag, Publ., Stuttgart, pp. 1–12.
Mazur, P.: 1977a, *Cryobiol.* 14, 251.
Mazur, P.: 1977b, in K. Elliott and J. Whelan (eds.), *The Freezing of Mammalian Embryos*, Ciba Foundation Symposium No. 52 (New Series), Elsevier, Amsterdam, pp. 19–42.
Mazur, P. and Schmidt, J.: 1968, *Cryobiol.* 5, 1.
Mazur, P., Farrant, J., Leibo, S. P., and Chu, E. H. Y.: 1969, *Cryobiol.* 6, 1.
Mazur, P., Barghoorn, E. S., Halvorson, H. O., Jukes, T. H., Kaplan, I. R., and Margulis, L.: 1978, *Space Sci. Rev.* 22, 3.
Michener, H. D. and Elliott, R. P.: 1964, *Adv. Food Res.* 13, 349.
Miller, R. H. and Mazur, P.: 1976, *Cryobiol.* 13, 404.
Morelli, F., Fehlner, F. P., and Stembridge, C. H.: 1962, *Nature* 196, 106.
Morris, G. J. and Farrant, J.: 1972, *Cryobiol.* 9, 173.
Nei, T., Araki, T., and Souzu, H.: 1965, *Cryobiol.* 2, 68.
Portner, D., Spiner, D. R., Hoffman, R. K., and Phillips, C. R.: 1961, *Science* 134, 2047.
Rall, W. F., Mazur, P., and Souzu, H.: 1978, *Biophys. J.* 23, 101.
Rapatz, G. L., Menz, L. J., and Luyet, B. J.: 1966, in H. T. Meryman (ed.), *Cryobiology*, Academic Press, N.Y., pp. 139–162.
Rapatz, G., Sullivan, J. J., and Luyet, B.: 1968, *Cryobiol.* 5, 18.
Rice, F. O.: 1960, in A. M. Bass and H. P. Broida (eds.), *Formation and Trapping of Free Radicals*, Academic Press, N.Y., p. 7.
Robinson, R. A. and Stokes, R. H.: 1959, *Electrolyte Solutions*, Academic Press, Inc., London.
Rose, A. H.: 1976, in *The Survival of Vegetative Microbes*, Symp. 26, Soc. Gen. Microbiol., Cambridge Univ. Press, pp. 155–182.
Sakurada, K.: 1958, *Low Temp. Sci. (Sapporo)* B16, 91.
Sant, R. K. and Peterson, W. H.: 1958, *Food Technology* 12, 359.
Scholander, P. F. and Maggert, J. E.: 1971, *Cryobiol.* 8, 371.
Scholander, P. F., van Dam, L., Kanwisher, J. W., Hammel, H. T., and Gordon, M. S.: 1957, *J. Cell. Comp. Physiol.* 49, 5.
Servin-Massieu, M. and Cruz-Camarillo, R.: 1969, *Appl. Microbiol.* 18, 689.
Smith, A. U. and Polge, C.: 1950, *Nature* 166, 668.
Souzu, H. and Mazur, P.: 1978, *Biophys. J.* 23, 89.
Souzu, H., Nei, T., and Bito, M.: 1961, *Low Temp. Sci.* B19, 49.
Webb, S. J.: 1960, *Can. J. Microbiol.* 6, 89.
Webb, S. J.: 1967a, *Can. J. Microbiol.* 13, 733.
Webb, S. J.: 1967b, *Nature* 213, 1137.
Whittingham, D. G., Leibo, S. P., and Mazur, P.: 1972, *Science* 178, 411.
Wood, T. H. and Rosenberg, A. M.: 1957, *Biochim. Biophys. Acta* 25, 78.
Wood, T. H. and Taylor, A. L.: 1957, *Radiation Res.* 6, 611.
Zentner, R. J.: 1966, *Bacteriol. Rev.* 30, 551.

BIOLOGICAL LIMITS OF TEMPERATURE AND PRESSURE

RICHARD Y. MORITA

Department of Microbiology and School of Oceanography, Oregon State University, Corvallis, OR 97331, U.S.A.

Abstract. Most biologists do not take into account that the greatest portion of today's biosphere is in the realm of environmental extremes, most of it being cold and under pressure. Since bacteria have the ability to adapt to environmental extremes, a close examination for the presence and/or growth of bacteria at high and low temperatures, low temperature and reduced pressure (less than 1 atm), low temperature and increased hydrostatic pressure should be made. It is also within the realm of possibility that life may have arisen in an environmental extreme on the primordial earth and then evolved over time to live under moderate temperatures and 1 atm. Microbial life has been demonstrated at temperatures slightly greater than 90 °C, below 0 °C, at hydrostatic pressures of 1100 atm, and possibly at cold temperatures in the atmosphere (less than 1 atm). Laboratory experiments have shown that certain enzyme reactions can occur above 100 °C under hydrostatic pressure, at −26 °C and at 5 °C under hydrostatic pressure.

1. Introduction

In reviewing the literature on the biological limits of temperature and pressure, the distinction must be made as to whether the organisms are actually growing and reproducing or merely surviving in the extreme environment. Many organisms have the ability to survive environmental extremes but do not necessarily grow or multiply in the extreme environment.

A difficulty encountered when trying to distinguish between survival and growth in nature is frequent lack of information concerning generation times. In the deep sea, for example, where energy sources are scarce, the temperature is around 5 °C or less and the hydrostatic pressure around 1,000 atm, generation times may be extraordinarily slow. Microbiologists tend to think of generation times of days and weeks characteristic of various cultures. For example, a clam (*Tindaria callistiformis*) obtained from a depth

TABLE I
Habitats showing large deviations from atmospheric pressures and mean temperatures

	Habitats
Hot temperatures (close to boiling point of water)	Hot springs, geysers
Low temperatures	Polar, high altitude regions
Hot temperature − high hydrostatic pressure	Red Sea thermal areas
Low temperature − low atmospheric pressure	Upper atmosphere
Low temperature − elevated atmospheric pressure	Caves
High temperature − high pressure	Oil deposits

Copyright © 1980 by D. Reidel Publishing Co., Dordrecht, Holland, and Boston, U.S.A.

of 3800 m took approximately 50–60 years for gonad development and 100 years to reach 8.4 mm size (Turekian *et al.*, 1975). Kuru provides another example of a slow virus which takes many years to express itself.

Many procaryotic cells may grow very slowly. Unfortunately, no studies of procaryotic cells in nature with generation times of weeks, or longer, have been undertaken.

Table I lists natural extreme pressure and/or temperature environments from which bacteria have been isolated.

2. High Temperature Environments

Thermophilic bacteria are organisms that grow optimally at temperatures of 50 °C or higher. Microbial survival and growth at hot temperatures have recently been reviewed by Tansey and Brock (1978), Amelunxen and Murdock (1978), and Brock (1978).

Natural hot environments are listed in Table II. Although the growth of organisms in erupting volcanoes has not been reported, bacteria have been observed in fumaroles at temperatures above the boiling point of water. Whether or not these microbes are surviving or growing is not known (Brock, 1967).

Both heterotrophic and chemolithotrophic species of bacteria grow at temperatures greater than 90 °C (Tansey and Brock, 1978). Photo-synthetic bacteria, including cyanobacteria have lower temperature maxima for growth (70 to 73 °C). The temperature limit for growth of eukaryotic organisms tends to be lower (60 to 62 °C) than for prokaryotes (Tansey and Brock, 1972). Brock (1967) concluded on the basis of his extensive experience with thermophilic bacteria in nature that "life is possible at any temperature at which there is liquid water".

Although certainly some proteins are more thermostabile than others and membrane stability plays an important role in thermophilic functions, the molecular basis for thermophily is far from understood (Amelunxen and Murdock, 1978).

3. Low Temperature Environments

Approximately 14% of the Earth's surface lies within polar regions. Most of the volume of the ocean is cold (5 °C or less), therefore, the major portion of the Earth's biosphere is cold.

TABLE II
Natural Hot Environments on Earth (from Tansey and Brock, 1978)

Erupting volcanoes: Temperatures up to 1000 °C.
Dry-steam fumaroles from volcanoes: Temperatures up to 500 °C.
Boiling and superheated springs: Temperatures from 93 to 101 °C, depending on altitude.
Non-boiling hot springs: Temperatures near boiling to ambient.
Sun-heated substrates (soils, litter, rock): Temperature depends on color and heat capacity.
  Temperatures reach 60 to 70 °C or higher.
Self-heating organic-rich materials (compost piles, seaweed piles, and coal refuse piles): Temperatures
  commonly up to 70 °C, and if ignition occurs above 100 °C.

TABLE III
Minimum temperature for growth of bacteria

Organism	Minimum temperature	Reference
5 isolates	−5.0	Horowitz-Wlassowa and Grinberg (1933)
10 isolates	−7.5	Bedford (1933)
Bacteria in ice cream	−10	Weinzirl and Gerdeman (1929)
Bacteria on fish	−11	Redford (1932)
Sporotrichum carnis	−7.5 to −10	Haines (1932)
Choetostylum fresenii and *Horomodendron cladosporoides*	−10	Bidault (1921)
Pseudomonads and molds in concentrated fruit and sugar solution	−18 to −20	Borgstrom (1961) Borgstrom (1961)
Aspergillus in glycerol	−18	Borgstrom (1961)
Pink yeasts on oysters	−18 to −30	McCormack (1950)

Psychrophilic bacteria are those with an optimum temperature for growth at 15 °C or lower, a maximum temperature for growth at 20 °C or lower and a minimum temperature for growth at 0 °C or lower (Morita, 1975). Psychrotropic bacteria are those organisms that are capable of growth at 5 °C (Eddy, 1960) or lower although their maximum growth temperature and optimum growth temperatures may be above the psychrophilic range. The older literature is marred by reports of growth at low temperatures which require confirmation by better techniques (Table III). A distinction must be made between growth and survival in such studies. Bacterial cultures are routinely preserved at low temperatures, indicating organisms can survive but not reproduce at low temperatures.

Growth of psychrophilic bacteria isolated from Arctic and Antarctic waters and maintained (0 °C or lower) at low temperatures have been studied recently by Morita (1975), Gillespie *et al.* (1976), Morita *et al.* (1977), Griffiths *et al.* (1977, 1978), and Baross and Morita (1978).

However, it should be noted that most psychrophilic bacteria must be maintained in permanently cold environments, temperature fluxations, even only to 10 °C or 15 °C may induce the death of these bacteria.

Lakes Vanda and Bonney, oligotrophic Arctic and Antarctic lakes respectively, are permanently frozen but have brackish bottom water. Photosynthetic activity takes place just below the ice layer (Goldman *et al.*, 1967). In Lake Bonney, Koob and Leister (1972) found culturable bacterial counts in the range of 6×10^5 l^{-1} at 8 to 9 m and 12 to 15 m. The temperature range in Lake Bonney was found to be from −2 °C beneath the ice to 7 °C. The highest bacterial counts were from samples taken from the 5 m layer under the ice incubated at 0 °C (Benoit *et al.*, 1971). The organisms were mainly yeast and an unusual microflora not readily cultured.

The relative humidity in the Dry Valley in Southern Victorialand, Antarctica is about 10 °C. In hour-long *in situ* experiments soil microbial activity was demonstrated, but the temperature was not mentioned (Vishniac and Mainzar, 1972). The one-hour interval was dictated by the constraints imposed on the investigators at sub-zero temperatures. The reproduction of bacteria took place only when liquid water which was provided by snowfall or melting snow was available.

Further information on life in Antarctic regions and on cold temperature microbes is provided by Baross and Morita (1978), and Inniss and Ingraham (1978).

4. High Temperature – High Hydrostatic Pressure Environments

High temperature ecosystems are not common: they are limited to the hot brine from the Atlantis II Deep of the Red Sea with temperatures up to 56 °C whereas the Discover Deep is about 45 °C. From these areas Truper et al. (1968) isolated *Desulfovibrio* spp. All isolates grew at 40 °C while two isolates showed a pronounced ability to grow at high temperatures and salinities. One isolate was able to survive in 10–17% NaCl incubated at 35 to 44 °C. Unfortunately, no pressure studies on these isolates were attempted.

Although thermophiles have been isolated from marine sediments (Bartholomew and Rittenberg, 1949; Bartholomew and Paik, 1966), they are probably inactive at the temperatures from which they were isolated.

5. High Temperature – High Pressure Environments

The temperature may range from 60 to 150 °C and the hydrostatic pressure from 200 to 400 atm in subterranean deposits. Thermophilic *Desulfovibrio* growing at 1000 atm and 104 °C was isolated by ZoBell (1958). Hydrogen sulfide also appeared to be produced by these organisms (ZoBell, 1958). *Desulfovibrio thermophilis* capable of growth at a temperature of 85 °C has also been isolated (Rozanova and Khudyakova, 1974).

Elevated pressures enable many common species of terrestrial bacteria to grow at temperatures above their normal maxima (ZoBell and Johnson, 1949). Increases in pressure can counteract increases in temperature as predicted by the Ideal Gas Law. Malic dehydrogenase functioning at 101 °C and inorganic pyrophosphatase functioning at 105 °C under increased hydrostatic pressure have been described (Morita and Haight, 1962; Morita and Mathemeier, 1964).

6. Low Temperature – Atmospheric Pressure Environment

Very few studies of low temperature habitats at atmospheric pressure have conducted. Although the atmospheric pressure may not be very high, this unique ecosystem is restricted mainly to subterranean caves. Microbiology of subterranean and glaciated caves of the Arctic, Lapland, Pyrenees and the Alps where the temperature range was 10 °C to below freezing have been conducted by Gounot (1967, 1968, 1969; Gounot et al., 1970). Although mainly psychrotropic bacteria were found only one true psychrophilic specie of bacterium was isolated. Only psychrotropic bacteria were isolated from the Karst caves of

Southern Indiana (Brock et al., 1973). The rate of growth was probably very slow due to the lack of energy (organic matter) present in the system.

7. Low Temperature – Low Atmospheric Pressure

At altitudes greater than 1000 m the temperature is generally 10 °C or lower, and the temperature and atmospheric pressure decreases further at greater altitudes. Proctor (1935) found that the isolates obtained from an altitude greater than 7000 m were typical soil bacteria. However, Proctor and Parker (1942) were able to grow 36 out of 105 isolated bacteria from an altitude greater than 3000 m at 0 °C. Fulton (1966) found bacteria at 3000 m (greater than 500 bacteria m^{-3}) over the polar regions where the temperature was −20 °C. Gregory (1961) and Burch (1967) demonstrated the presence of microorganisms at 10 000 m (troposphere, −40 °C) and 27 000 m (stratosphere), respectively.

In 1961 Gregory proposed a 'biological zone' where bacteria grew on organic enriched droplets of water and fixed nitrogen. Parker and Watchtel (1971) indicated that bacteria could grow in clouds which would enrich the rain water. Relatively high levels of vitamins (cobalamin, biotin, and niacin) presumed to be of bacterial origin were found in rain water (Parker and Wachtel, 1971).

8. Low Temperature – Hydrostatic Pressure Environments

The deep sea environment is characterized by low temperature and hydrostatic pressure. Below the thermocline the temperature remains rather constant, being 5 °C or lower, whereas the hydrostatic pressure increases 1 atm per 10 m in depth. Psychrophilic bacteria are not present in abundance in the deep sea (Morita, 1976; Norkrans and Stehn, 1978). The open ocean, especially the deep sea, does not have a high organic matter content. The particulate organic matter ranges from 30 to 10 μg carbon l^{-1}, and the dissolved organic carbon ranges from 0.35 to 0.7 mg carbon l^{-1} (Menzel and Ryther, 1970).

Although bacteria have been isolated from all depths of the ocean and oceanic sediment; growth and metabolism of deep sea organisms are slow. Enrichment cultures obtained from the Weber Trench (collected during the Galathea Deep-Sea Expedition from a depth of 7250 m) had to be incubated nearly one year at 5 °C and 700 atm in order to obtain a positive indication of sulfate reduction (Morita, 1976).

An illustration of metabolic rate *in situ* was given by Jannasch and Wirsen (1973) who subjected agar and other substrates to the indigenous microflora and incubated them for one year at 4 °C at a depth of 1830 m. During this period 1.3 and 1.7% (duplicates) of the agar were utilized and the substrate concentration was 0.33 mg agar ml^{-1}. The lower limit of 1.3% gives a value of 0.0043 mg ml^{-1} of agar utilized. Data from Broecker (1963) for the upper North Atlantic Deep Water, which was approximately in the same region as the incubation site of Jannasch and Wirsen (1973), indicates that this water has an apparent age or residence time of 360 ±160 years.

An extrapolation from Jannasch and Wirsen's (1973) shows the amount of agar utilized in 350 years would be 1.505 g l^{-1} with a lower limit of 817 mg of agar per liter utilized in 190 years. The data from Menzel and Ryther (1970) indicate that the particulate organic carbon in the open ocean is 3 to 10 μg C l^{-1}, and the dissolved organic carbon is 0.35 to 0.70 mg C l^{-1}. Hence, the extrapolated values for Jannasch and Wirsen's data are much too high. Two conclusions can be drawn. Surface effects of the bottles used for the experiments and the addition of substrate will have an effect on the result. Both of these factors may increase the metabolic rate of the indigenous microflora.

Difficulties occur when one tries to distinguish between survival and slow metabolic rate. We know that organisms can survive in the deep sea. Starvation increases barotolerance (Novitsky and Morita, 1978a) resulting in survival. Furthermore, there appears to be a survival mechanism for a marine vibrio, where large numbers of cells are produced with no increase in biomass, when a cell is subjected to starvation (Novitsky and Morita, 1978b).

Microbes in the deep sea environments are discussed in the reviews by Marquis and Matsumura (1978) and Morita (1967, 1972, 1973).

9. Conclusion

Although much work has been done on the ability of thermophiles to live at high temperatures, the mechanisms by which microorganisms live in extreme environments of temperature and pressure have not yet been fully explained.

The ability of organisms to survive has not yet been clearly distinguished from growth in extreme environments. Very slow metabolic rates may occur in non-optimal environments, particularly when the environment lacks oxidizable organic matter or compounds such as reduced sulfur as an energy source. What we think may be a depressed metabolism may actually be a metabolic rate in harmony with the environment.

Simple survival permits the microbial genome to be preserved until the right environment is present. We have become accustomed to rapid metabolism in organisms such as *Escherichia coli*, where these values can be measured. Most organisms however, living in soil or the aquatic environment, probably have mechanisms of survival with metabolic rates that are not easily measurable. Generally, bacteria undergo spore formation to survive adverse environments, and the metabolic rate of these spores is not measurable. The question of survival versus slow metabolic rates arises.

Life may have originated in a harsh environment where certain inorganic chemical reactions can occur more readily. In such a situation, the metabolic rate of the organism may have been very slow, occurring over a long span of time. When conditions became more optimal, organisms evolved faster metabolic rates. The availability of energy is one of the main factors that will determine the rate of metabolic processes in an organism.

References

Amelunxen, R. E. and Murdock, A. L.: 1978, in D. J. Kushner (ed.), *Microbial Life in Extreme Environments*, Academic Press, London, New York, and San Francisco, pp. 217–278.

Bartholomew, J. W. and Paik, G.: 1966, *J. Bacteriol.* **92**, 635.
Bartholomew, J. W. and Rittenberg, S. C.: 1949, *J. Bacteriol.* **57**, 659.
Baross, J. A. and Morita, R. Y.: 1978, in D. J. Kushner (ed.), *Microbial Life in Extreme Environments*, Academic Press, London, New York, and San Francisco, pp. 9–72.
Baross, J. A., Hanus, F. J., and Morita, R. Y.: 1975, *Appl. Microbiol.* **30**, 309.
Benoit, R., Hatcher, R., and Green, W.: 1971, in J. Cairns, Jr. (ed.), *The Structure and Function of Freshwater Microbial Communities*, Virginia Polytechnic Institute and State University, Blacksburg, Virginia, pp. 287–293.
Bedford, R. H.: 1933, *Contrib. Can. Biol. Fish.* **7**, 433.
Bidault, C.: 1921, *Compt. Rend. Soc. Biol.* **85**, 1017.
Borgstrom, G.: 1961, *Proc. Low Temp. Microbiol. Symp.*, Campbell Soup Co., pp. 197–250.
Broecker, W.: 1963, in M. N. Hill (ed.), *The Sea*. Vol. 2. Interscience Publ. Co. New York, London, and Sydney, pp. 88–127.
Brock, T. D.: 1967, *Science* **158**, 1012.
Brock, T. D.: 1978, *Thermophilic Microorganisms and Life at High Temperatures*, Springer-Verlag, New York.
Bruch, C. W.: 1967, in P. H. Gregory and J. L. Monteith (eds.), *Airborne Microbes*, Seventeenth Symp. Soc. Gen. Microbiol., Cambridge Univ. Press, London, pp. 354–374.
Eddy, B. P.: 1960, *J. Appl. Bacteriol.* **23**, 189.
Fulton, J. D.: 1966, *Appl. Microbiol.* **14**, 233.
Gillespie, P. A., Morita, R. Y., and Jones, L. P.: 1976, *J. Oceanogr. Soc. Japan* **32**, 74.
Griffiths, R. P., Hayasaka, S. S., McNamara, T. M., and Morita, R. Y.: 1977, *Appl. Environ. Microbiol.* **34**, 801.
Griffiths, R. P., Hayasaka, S. S., McNamara, T. M., and Morita, R. Y.: 1978, *Can. J. Microbiol.* **24**, 1217.
Goldman, C. R., Mason, D. T., and Hobbie, J. E.: 1967, *Limnol. Oceanogr.* **12**, 295.
Gounot, A.-M.: 1967, *Ann. Inst. Pasteur.* **113**, 923.
Gounot, A.-M.: 1968, *Comp. Rendus Acad. Sci. Parks.* **D266**, 1437.
Gounot, A.-M.: 1969, *V. Int. Kongr. Spelanologie, Stuttgart* **4**, 1.
Gounot, A.-M., Breuil, C., Borgere, P., and Simeon, D.: 1970, *Compt. Rend. 9th Congr. Nat. Speleol. Dijon. Spelunca Mem.* **7**, 123.
Gregory, P. H.: 1961, *Microbiology of the Atmosphere*, Leonard Hill (Books) Ltd., London.
Haines, R. B.: 1931, *J. Exptl. Biol.* **8**, 379.
Horowitz-Wlassowa, L. M. and Grinberg, L. D.: 1933, *Zentr. Bakteriol. Parasitenk., Abt. 11* **89**, 54.
Inniss, W. E. and Ingraham, J. L.: 1978, in D. J. Kushner (ed.), *Microbial Life in Extreme Environments*, Academic Press, London, New York, and San Francisco, pp. 73–104.
Jannasch, H. W. and Wirsen, C. O.: 1973, *Science* **180**, 641.
Koob, D. D. and Leister, G. L.: in G. A. Llano (ed.), *Antarctic Terrestrial Biology*, Vol. 20, Antarctic Research Series, American Geophysical Union, Washington, D.C., pp. 51–68, 72.
Marquis, R. E. and Matsumura, P.: 1978, in D. J. Kushner (ed.), *Microbial Life in Extreme Environments*, Academic Press, London, New York, and San Francisco, pp. 105–158.
McCormack, G.: 1950, *Comm. Fish. Rev.* **12**, 28.
Menzel, D. W. and Ryther, J. H.: 1970, in D. W. Hood (ed.), *Symposium on Organic Matter in Natural Waters*, Inst. Mar. Sci. Occasional Publ. No. 1, Univ. Alaska, Fairbanks, pp. 31–54.
Morita, R. Y.: 1967, *Oceanogr. Mar. Biol. Ann. Rev.* **5**, 187.
Morita, R. Y.: 1972, in O. Kinne (ed.), *Marine Biology – Environmental Factors: A Treatise*, Inter-Science Publ. Co., London, pp. 1361–1388.
Morita, R. Y.: 1973, in R. Bauer (ed.), *Barobiology and Experimental Biology in the Deep-Sea*, Jniv. N. Carolina Press. Chapel Hill, NC, pp. 89–105.
Morita, R. Y.: 1975, *Bacteriol. Rev.* **39**, 144.
Morita, R. Y.: 1976, in T. G. R. Gray and J. R. Postgate (eds.), *The Survival of Vegetative Microbes*, Twenty-Sixth Symp. Soc. Gen. Microbiol., Cambridge Univ. Press, Cambridge, pp. 279–298.
Morita, R. Y. and Haight, R. D.: 1962, *J. Bacteriol.* **83**, 1341.
Morita, R. Y. and Mathemeier, P. F.: 1964, *J. Bacteriol.* **88**, 1667.
Morita, R. Y., Griffiths, R. P., and Hayasaka, S. S.: 1977, in G. A. Llano (ed.), *Adaptations within*

Antarctic Ecosystems, Proceedings of the Third SCAR Symposium on Antarctic Biology, Gulf Publ. Co., Houston, pp. 99–113.
Novitsky, J. A. and Morita, R. Y.: 1978a, *Mar. Biol.* 48, 289.
Novitsky, J. A. and Morita, R. Y.: 1978b, *Mar. Biol.* 49, 7.
Norkrans, B. and Stehn, B. O.: 1978, *Mar. Biol.* 47, 201.
Parker, B. C. and Wachtel, M. A.: 1971, in J. Cairns, Jr. (ed.), *The Structure and Function of Fresh-Water Microbial Communities*, Res. Div. Mon. 3, Virginia Polytechnic Institute and State University, Blacksburg, Va., pp. 195–207.
Proctor, B. E.: 1935, *J. Bacteriol.* 30, 363.
Procter, B. E. and Parker, B. W.: 1942, in S. Moulton (ed.), *Aerobiology*, AAAS Publ. No. 17, Washington, D.C., pp. 48–54.
Redfort, A. L.: 1932, *Bull. Intern. Renseigh. Frigoritiques* 4, 40.
Rozanova, E. O. and Khudyakova, A. I.: 1974, *Mikrobiologiya* 43, 1069.
Tansey, M. R. and Brock, T. D.: 1972, *Proc. Natl. Acad. Sci., U.S.A.* 69, 2426.
Tansey, M. R. and Brock, T. D.: 1978, in D. J. Kushner (ed.), *Microbial Life in Extreme Environments*. Academic Press, London, New York, and San Francisco, pp. 159–216.
Trüper, H. G., Kelleher, J. J., and Jannasch, H. W.: 1968, *Arch. Mikrobiol.* 65, 208.
Turekian, K. K., Cochran, J. K., Kharkar, D. P., Cerrato, R. M., Vaisnys, J. R. Sanders, H. L., Grassle, J. F., and Allen, J. A.: 1975, *Proc. Nat. Acad. Sci.*, U.S.A.
Turekian, K. K., Cochran, J. K., Kharkar, D. P., Cerrato, R. M., Vaisnys, J. R. Sanders, H. L., Grassle, J. F., and Allen, J. A.: 1975, *Proc. Nat. Acad. Sci., U.S.A.* 72, 2829.
Vishniac, W. V. and Mainzer, S. E.: 1973, *Antarctic J. U.S.* 7, 88.
Weinzirl, J. and Gerdeman, A. E.: 1929, *J. Dairy Sci.* 12, 182.
ZoBell, C. E.: 1958, *Producers Monthly* 22, 12.
ZoBell, C. E. and Johnson, F. H.: 1949, *J. Bacteriol.* 57, 179.

ENDOLITHIC MICROBIAL LIFE IN HOT AND COLD DESERTS

E. IMRE FRIEDMANN

Department of Biological Science, Florida State University, Tallahassee, Fl 32306, U.S.A.

Abstract. Endolithic microorganisms (those living inside rocks) occur in hot and cold deserts and exist under extreme environmental conditions. These conditions are discussed on a comparative basis. Quantitative estimates of biomass are comparable in hot and cold deserts. Despite the obvious differences between the hot and cold desert environment, survival strategies show some common features. These endolithic organisms are able to 'switch' rapidly their metabolic activities on and off in response to changes in the environment. Conditions in hot deserts impose a more severe environmental stress on the organisms than in the cold Antarctic desert. This is reflected in the composition of the microbial flora which in hot desert rocks consist entirely of prokaryotic microorganisms, while under cold desert conditions eukaryotes predominate.

Microorganisms living inside rocks in hot or cold deserts exist in severe environments characterized by low humidity, extreme temperatures and considerable daily temperature variations. They also may be exposed to high levels of solar radiation.

Despite the common physiographic features shared by rocky hot deserts and cold polar deserts, life in these places appears in very different forms. In most hot deserts, the presence of plants and animals is a conspicuous feature many of which often show spectacular adaptations to the desert environment. In contrast, the dry valleys of Southern Victoria Land, Antarctica, probably the most extreme polar desert on Earth, appear completely lifeless. Some views of the dry valleys are reminiscent of the Martian landscape, by now familiar from photographs of the Viking mission. This similarity perhaps goes beyond superficial likeness, the extremes of cold and drought in the dry valleys result in an environment that has been regarded as the closest terrestrial analog of the Martian surface.

Microbiological studies of dry valley soils yielded somewhat conflicting results. Although the presence of microorganisms was demonstrated by culture methods, there are indications that these may not have been indigenous but were carried in by winds from more temperate climates (Horowitz *et al.*, 1972). Other studies found actively growing bacteria, algae and yeasts in dry valley soils (Uydess and Vishniac, 1976, Vishniac and Hempfling, 1979).

Our studies were oriented towards microorganisms that inhabit rock substrates rather than soils. Rocky landscapes, such as those shown in Figures 1 and 2 occur in both hot and cold, polar deserts. Beneath the bare surface of the rocks, a rich endolithic microbial growth may be present, in morphological appearance similar in both hot and cold deserts. Microorganisms may colonize minute cracks in weathering rocks and form the so-called

Fig. 1. Landscape with Nubian sandstone cliffs near Timna, Negev desert, Israel.

Fig. 2. Landscape with Beacon sandstone cliffs. The dark intrusion is dolerite. University valley (third lateral valley of Beacon valley), Southern Victoria Land, Antarctica.

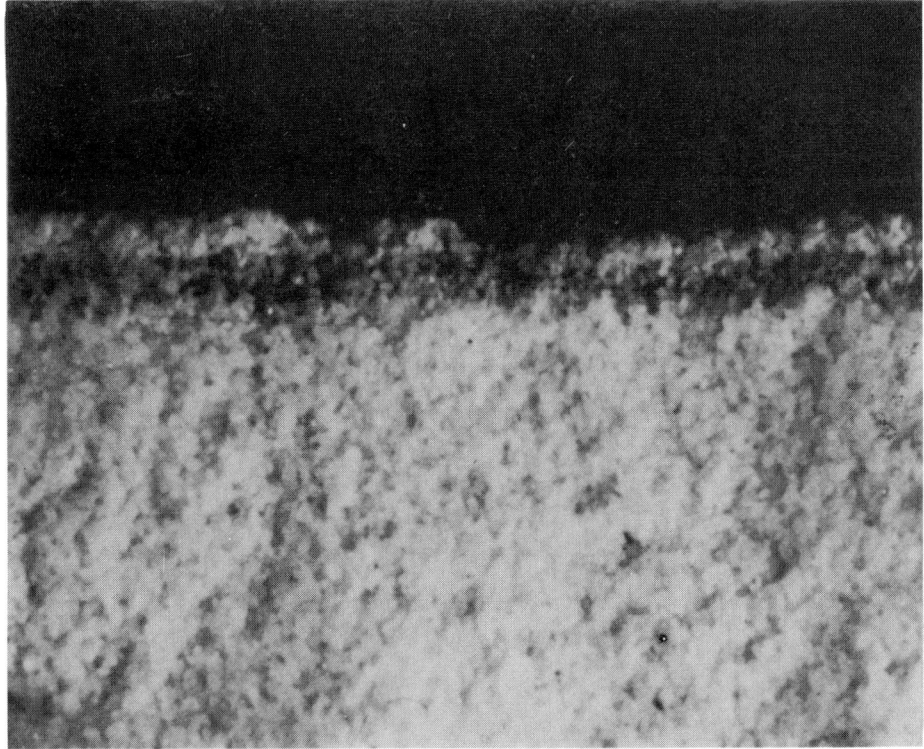

Fig. 3. Fractured sandstone with dark zone of cryptoendolithic blue-green algae (cyanobacteria) under the surface. Sinai desert. Magnification: ×3.5.

chasmoendolithic growth. A more conspicuous growth form exists in certain porous rocks such as porous sandstone. Here microorganisms form a thin green or dark brown layer underneath and parallel to the rock surface. This growth form, termed cryptoendolithic, occurs in both hot and cold polar deserts (Friedmann, 1972, 1977; Friedmann and Galun, 1974; Friedmann et al., 1967; Friedmann and Ocampo, 1976). The appearance of the endolithic community from the Sinai desert (Figure 3) can be compared to that of the Antarctic dry valleys (Figure 4). In both cases, the organisms occupy a definite colored zone a few millimeters thick under the surface. The spatial arrangement within the porous rock and the attachment of microorganisms to the substrate can be seen in the scanning electron micrograph shown in Figure 5.

How do cryptoendolithic organisms survive in the harsh hot and cold desert environments? The SEM micrographs of a fractured sandstone shown in Figures 6 and 7 offer some clue. This rock, from the Negev desert in Israel, is colonized by endolithic cyanobacteria between the levels marked by arrows (individual cells as shown in Figure 5 are not visible at this magnification). The rock has a distinct porous structure but near the surface the airspaces between the rock particles are filled with inorganic debris that forms a solid crust. When photographed from the surface (Figure 7), the crust appears continuous,

Fig. 4. Fractured sandstone with dark zone of cryptoendolithic blue-green algae (cyanobacteria) under the surface. Asgard range, Southern Victoria Land, Antarctica. Magnification: x3.5.

lacking spaces between the particles. The crust, although permeable to water and gases, is a barrier to the penetration of organisms, and the cryptoendolithic microenvironment that exists within the airspace system of the porous rock is separated from the outside environment. Consequently, the microclimate inside the rock can be distinctly different from the climate exterior to the rock.

A brief survey of the physical factors that affect the cryptoendolithic environment may be of interest. A major factor is light. Endolithic microbial communities in desert rocks always seem to contain a primary producer. These photosynthetic organisms utilize the sunlight that penetrates the upper few millimeters of the rock. Consequently, only those rock types that are light colored and to some extent translucent, can be colonized by cryptoendolithic microorganisms, as the availability of light is a condition of endolithic microbial life. There is a steep light intensity gradient within the upper few millimeters of the rock that is the likely cause of the sharp limits of the cryptoendolithic zone. This assumption is further supported by the observation that cells in the upper level of the zone often have walls with dark pigments that are likely to function as a light-protective filter. At the same time, cells in the lowest level have colorless walls and are rich in photosynthetic pigments in an apparent response to low light intensity.

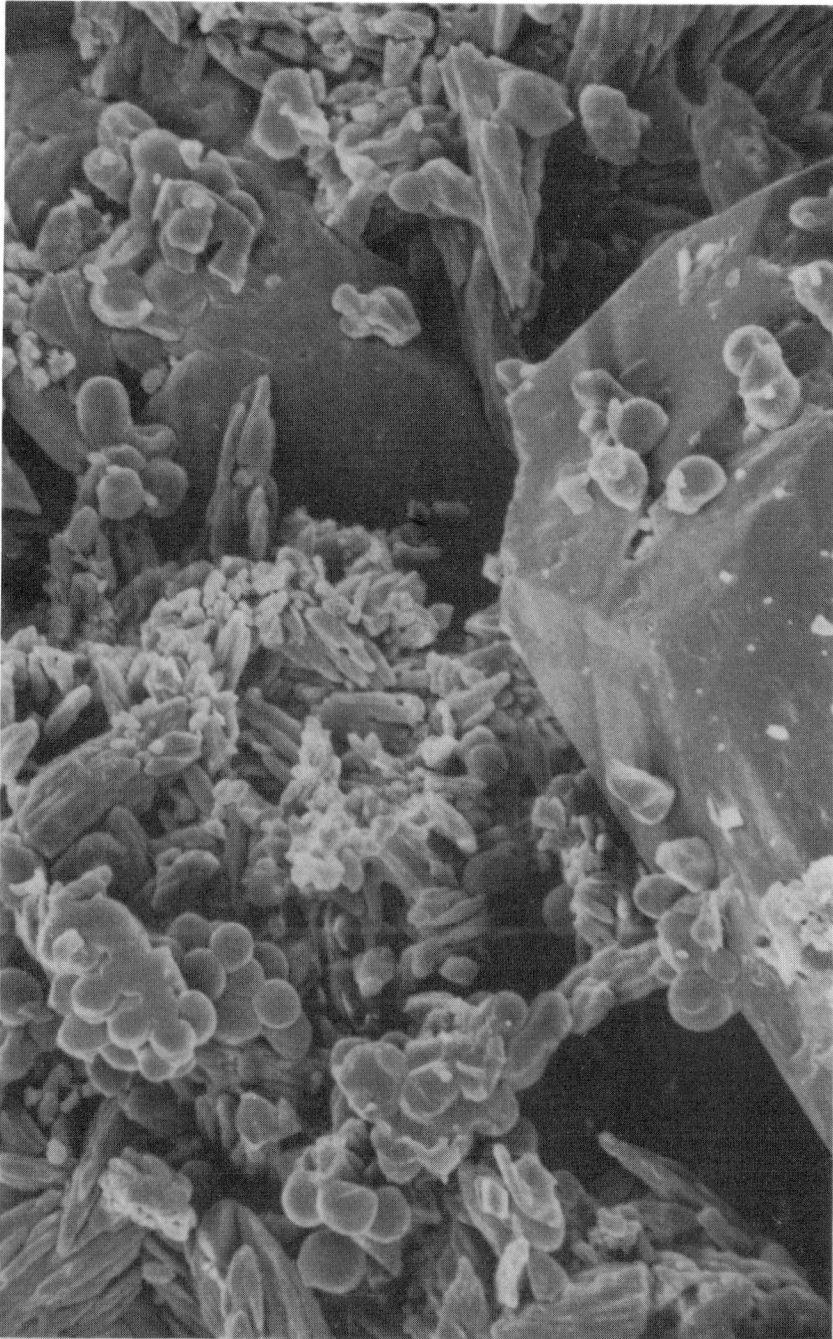

Fig. 5. Scanning electron micrograph of fractured surface of the cryptoendolithic microbial zone in sandstone, showing large hexagonal quartz crystals, small plate-like matrix crystals and spherical cells of a blue-green alga. Negev desert, Israel. Magnification: ×2000. (After Friedmann, 1971).

Fig. 6. Low magnification scanning electron micrograph of fractured surface of sandstone, Negev desert, Israel. Between the levels indicated by arrows the rock is colonized by cryptoendolithic microorganisms, although the cells are not visible at this magnification. The airspace system between the rock particles is apparent. Near the surface, the spaces between the particles are filled with mineral substance. Magnification: x60. (After Friedmann, 1971).

Fig. 7. Scanning electron micrograph of the surface crust of the rock shown in Figure 6. The spaces between the rock particles are filled by dense mineral substance. Magnification: x60. (After Friedmann, 1971).

Insolation also affects the temperature of the rock, although conditions in hot and cold deserts differ in this respect. In hot deserts, the temperature of rock surfaces reached by solar radiation can rise considerably above that of the air, while at night, it sinks to ambient level. Figure 8 shows temperature variations at the rock surface and inside the rock during a typical day in the Sonoran desert of Mexico.

In the Antarctic cold desert, temperature conditions are more complex. During the polar winter, there is no solar radiation, and temperatures may drop to −60 °C or lower. During the summer, ambient air temperatures are generally between −15 °C and 0 °C. Because of the intense solar radiation, rock temperatures can rise considerably higher, with a correlation between exposure of the surface, temperature and microbial colonisation. Rock surfaces with a northern exposure receive prolonged direct insolation, and rock temperatures about 20 °C above the ambient value are common. At the same time, the temperature of rock surfaces exposed to the South do not rise much above that of the air. Figure 9 illustrates the temperature changes at the northern and at the southern faces of a sandstone boulder, in the dry valleys of Southern Victoria Land, during a summer day. Only the northern, 'warm' surface of the rock is colonized by microorganisms while the southern, 'cold' surface remains abiotic.

Such conditions result in a rather conspicuous phenomenon. In the dry valleys, the

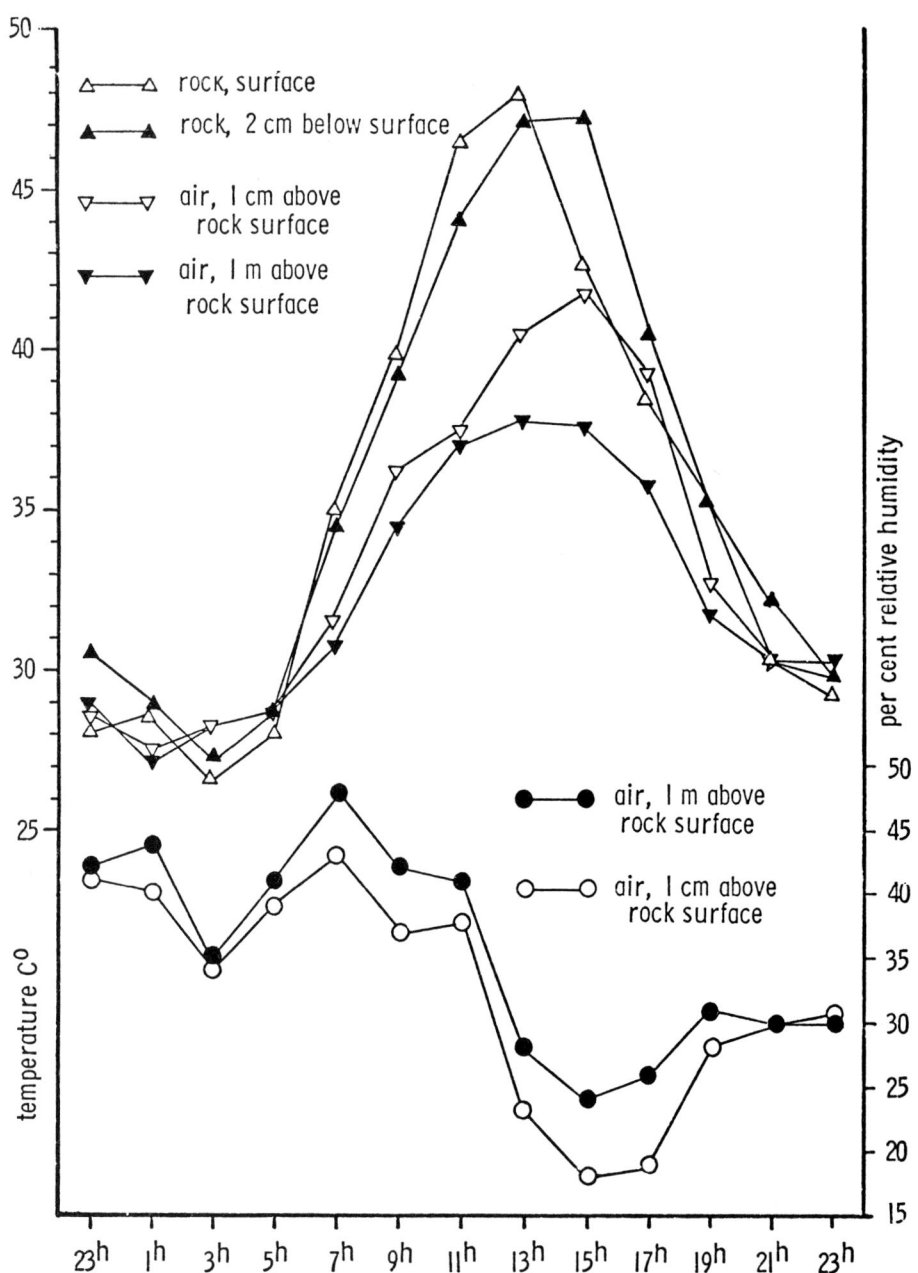

Fig. 8. Daily variations in relative humidity and in temperature of air, and of granite rock colonized by chasmoendolithic microorganisms. June 23, 23^h to June 24, 23^h, 1971. Sonoran desert, Mexico. 32° 04′ N, 113° 14′ W. (After Friedmann, 1972).

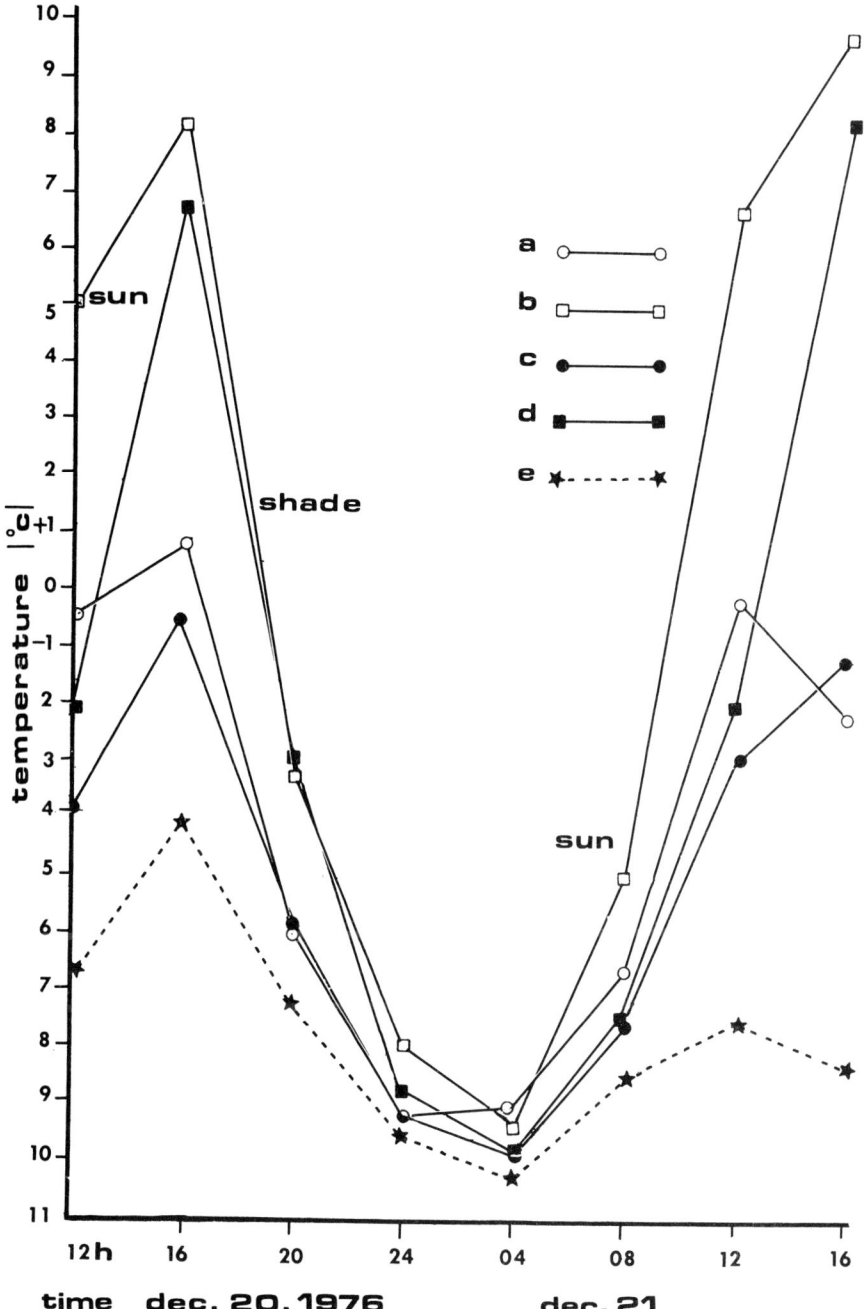

Fig. 9. Daily variations in temperature of air and of Beacon sandstone rock. 'Warm' face of rock (with cryptoendolithic lichen) facing NW: (a) surface of rock, (b) 3 centimeters deep in the rock. 'Cold' face of rock facing SW (no lichens), (c) surface of rock, (d) 3 centimeters deep in the rock, (e) air. Tyrol valley, Asgard range, Southern Victoria Land, Antarctica. 77° 35' S, 160° 39' E. (After Friedmann, 1977).

Fig. 10. 'Warm' face of sandstone boulder, facing North, Beacon valley, Southern Victoria Land. The exfoliating weathering pattern is an indication of colonization by cryptoendolithic lichen. Measuring bar: 50 cm.

most common type of microbial colonization produces a typical exfoliating weathering pattern on the surface of sandstone (Figure 10). At the northern 'warm' face of sandstone rocks, this pattern, indicative of microbial colonization under the surface, is mostly apparent. In contrast, the southern 'cold' face of the rocks is smooth, due to wind erosion.

Water is evidently an important environmental factor in deserts with condensation of dew as the principle source of water for endolithic microorganisms in most, or perhaps in all hot deserts (Friedmann, 1971; Friedmann and Galun, 1974; Friedmann *et al.*, 1967). In hot deserts dew is more abundant than rain and generally is more evenly distributed throughout the year. In the Antarctic dry valleys, melting of snow seems to be the principle, or perhaps only source of water (Friedmann, 1978).

The water content of sandstone rocks colonized by endolithic microorganisms was determined in several samples from the dry valleys of Antarctica, from the American desert and from the Sinai peninsula. The numerical values of these measurements were remarkably similar, amounting to about 0.1% to 0.2% of the weight of the rocks. This quantity of moisture is probably sufficient to maintain a fully or nearly saturated atmosphere in the microscopic air space system of cold desert rocks. The situation is different in hot deserts as at high temperatures the water that is present inside the rocks results in very low relative humidity values. This is also evident from relative humidity measurements (Figure 7). When temperatures of insolated rocks rose above the ambient level, the relative humidity of the air near the rock surface dropped dramatically.

Although cryptoendolithic microbial modes of growth in hot and cold desert rocks are morphologically quite similar (Figures 3 and 4), the types of organisms are different. In hot deserts, only prokaryotes exist in the cryptoendolithic habitat. The dominant organism is always a coccoid blue-green alga (cyanobacterium). Primarily only a single species is present, accompanied by some colorless bacteria. The Culture Collection of Xerophytic Microorganisms at Florida State University maintains over 150 strains of cyanobacteria isolated from cryptoendolithic growths in North and South America, Asia, and Africa. Nearly all of these are species of the genus *Chroococcidiopsis* or of related genera. This group of cyanobacteria seems to be particularly adapted to the cryptoendolithic environment.

In the Antarctic dry valleys, microbial colonization of rocks by coccoid cyanobacteria is rather infrequent, and the most common cryptoendolithic organism is an unusual lichen (Figure 11). In contrast to endolithic lichens in other climates, this organism exists entirely under the surface of the rock, growing around the rock particles in the sponge-like airspace system of the porous rock. The mycobiont (algal component) is probably an ascomycete, although generally it does not form fruiting bodies. The phycobiont (algal component) is mostly *Trebouxia* (Chlorophyta), a eukaryotic alga common in lichens. This cryptoendolithic lichen mostly forms characteristic colored zones under the surface of the rock. The uppermost zone is black because of the dark pigmented fungus hyphae that generally enclose phycobiont cells. The zone below is white, and colorless fungus hyphae are abundant. The lowest zone is green; this color resulting from a concentration of green phycobionts.

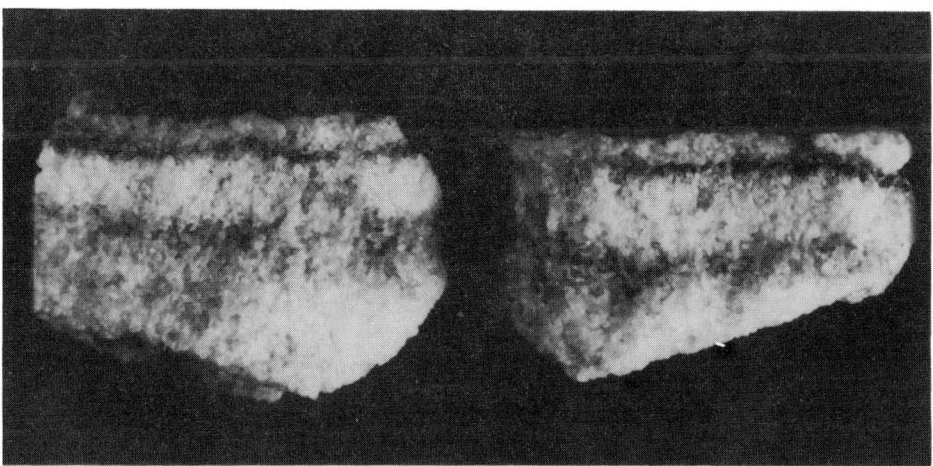

Fig. 11. Fractured sandstone with colored zones of cryptoendolithic lichen. Upper, black zone: dark pigmented fungus hyphae and phycobiont cells. Wider, white zone: Fungus hyphae with few phycobiont cells. Lower, green zone: Phycobionts with some fungus hyphae. Asgard range, Southern Victoria Land, Antarctica. Magnification: x3.5.

TABLE I

Nitrogen, ATP, chlorophyll and estimated total organic matter in 20 rock samples from the Antarctic dry valleys

Kjeldahl N mg m^{-2}	ATP mg m^{-2}	Chlorophyll mg m^{-2}	N/ATP	Chlorophyll/ ATP	Organic matter (N x 25) g m^{-2}
1.29–7.07	(0)0.8–15.7	1.7–24.9	(0)167–3210	(0)0.36–10.83	32.25–176.75

(Data after Friedmann et al., in press).

We attempted to make a quantitative estimate of organic matter in endolithic microbial communities, based on Kjeldahl nitrogen, ATP and chlorophyll fluorescence analyses (Table I). From the amount of nitrogen, the total organic matter in dry valley rocks is estimated as ranging between 32.25 to 176.75 g m^{-2} rock surface. Table I also shows a wide nitrogen/ATP ratio. High values may indicate the presence of dead organic matter in some samples. The variation in the chlorophyll/ATP ratio perhaps reflects differences in the ratio between photosynthetic organisms (algae) and non-photosynthetic fungi and bacteria. Comparative studies with hot desert endolithic growths gave remarkably similar results. In sandstone and granite rocks from the Mojave desert in California, the Sonoran desert in Mexico and the Negev in Israel, the total organic matter, estimated from Kjeldahl nitrogen determinations, ranged from 37.5 to 185.0 g m^{-2} rock surface.

How do these organisms cope with their harsh environment? In hot deserts and in the Antarctic dry valleys, cryptoendolithic microorganisms share a common survival strategy. They are capable of 'switching' their metabolic activities on and off in response to rapid changes in environmental conditions. In the Antarctic environment, during winter or in temporal absence of direct insolation, the organisms are inactive in a 'freeze-dried' state that apparently does not damage cellular structures. During warmer periods, temperatures inside the rock may rise to around 10 °C. At this temperature, even relatively small quantities of water can maintain a level of humidity that is sufficient for metabolic activity. Under these conditions it is not surprising that eukaryotic organisms such as green algae and fungi find accommodation in the Antarctic endolithic environment.

In contrast, in hot deserts the most favorable environmental combination of water availability and of moderate temperatures occurs in the early morning hours after dew fall. Later in the day, the temperature rises and relative humidity drops drastically. The combination of high temperature and extreme aridity imposes so severe an environmental stress that eukaryotic organisms are unable to survive. It is therefore, understandable why in hot deserts, endolithic microbial communities are composed of prokaryotes only and why the number of species is generally lower than it is in the Antarctic desert.

In conclusion, the endolithic environment in hot deserts is much harsher and more extreme than its counterpart in the Antarctic polar desert. This is so in spite of the fact that the outside environment in the dry valleys is more 'hostile' to life than it is in hot desert areas. The endolithic environment is an integral microscopic world upon which outside conditions have only limited influence.

Acknowledgements

I am grateful to my wife, Roseli Ocampo-Friedmann, who actively participated in every stage of this research. The support of the National Aeronautics and Space Administration (grant No. 7337) and of the National Science Foundation, Division of Polar Programs (grants DPP 76–1551 and DPP 77–21858) is thankfully acknowledged.

References

Friedmann, E. I.: 1971, *Phycologia* **10**, 411.
Friedmann, E. I.: 1972, in L. E. Rodin (ed.), *Ecophysiological Foundation of Ecosystems Productivity in Arid Zones*, Nauka, U.S.S.R. Academy of Sciences, Leningrad, pp. 182–185.
Friedmann, E. I.: 1977, *Antarct. J. U.S.* **12**, 26.
Friedmann, E. I.: 1978, *Antarct. J. U.S.* **13**, 162.
Friedmann, E. I. and Galun, M.: 1974, in G. W. Brown, Jr. (ed.), *Desert Biology*, Vol. II, Academic Press, New York and London, pp. 165–212.
Friedmann, E. I. and Ocampo, R.: 1976, *Science* **193**, 1247.
Friedmann, E. I., Lipkin, Y., and Ocampo-Paus, R.: 1967, *Phycologia* **6**, 185.
Friedmann, E. I., La Rock, P. A., and Brunson, J. O.: 1980, *Antarct. J.U.S.* **15**, (in press).
Horowitz, N. H., Cameron, R. E., and Hubbard, J. S.: 1972, *Science* **176**, 242.
Uydess, I. L. and Vishniac, W. V.: 1976, in M. R. Heinrich (ed.), *Extreme environments. Mechanisms of Microbial Adaptation*, Academic Press, New York, San Francisco, London, pp. 29–56.
Vishniac, H. L. and Hempfling, W. P.: 1979, *Int. J. Syst. Bacteriol.* **29**, 153.

PURIFICATION AND PROPERTIES OF MALATE DEHYDROGENASE FROM THE EXTREME THERMOPHILE *BACILLUS CALDOLYTICUS*

HORDUR KRISTJANSSON and CYRIL PONNAMPERUMA

Laboratory of Chemical Evolution, Department of Chemistry, University of Maryland, College Park, MD 20742, U.S.A.

Abstract. The enzyme malate dehydrogenase (EC 1.1.1.37) from an extreme thermophile *B. Caldolyticus* was purified to about 91% homogeneity. The molar mass of the enzyme was determined as 73 000 daltons and it is composed of two subunits, each with a molar mass of 37 000. Initial velocity studies with oxaloacetic acid and NADH as substrates at pH 8.1, over a range of temperatures, indicate that the enzyme operates via a sequential type mechanism. Van't Hoff plots of the kinetic parameters displayed sharp changes in slope at characteristic temperatures, whereas the Arrhenius plot exhibited no such breaks over the temperature interval investigated. The enzyme was found to be stable at 41 °C and lower temperatures. At 51 °C and 59 °C an almost immediate 20% reduction in activity was obtained, but no further inactivation occurred during the 60 min of incubation. At 59 °C the enzyme lost 50% of its initial activity in about 38 s. High concentration of NADH was observed to greatly stabilize the enzyme at that temperature.

It is suggested that the slope changes in the Van't Hoff plots and the stability profiles at 51 °C and 59 °C are representative of a temperature induced conformational change in the enzyme.

Adaptation of thermophilic microorganisms to a hostile environment allows them to survive and grow at temperatures close to the boiling point of water. The mechanisms of thermal tolerance are diverse, involving protective lipid interactions, rapid metabolic turnover, and the existence of inherently stable cell constituents in the organism. An important aspect of survival is the maintenance of functional enzymes at high temperatures. Studies have indicated, that although in most instances the thermophilic enzymes exhibit considerable stability over their mesophilic counterparts, properties such as molar mass, allosteric properties, amino acid composition, and even primary structure, are found to be amazingly similar between the enzymes from the two different sources (Kristjansson, 1977).

The present study was undertaken to investigate the adaption of thermophilic microorganisms to growth at higher temperatures. This was done through an investigation of the stability and kinetic properties of an enzyme that would seem important for survival and function. An extreme thermophile, *Bacillus caldolyticus* was selected for the extraction of the enzyme, in order to maximize its temperature dependent structural and behavioral differences, as compared to its mesophilic counterparts.

Malate dehydrogenase was purified to about 91% homogeneity, as determined by polyacrylamide gel electrophoresis. The methods utilized were a combination of ammonium sulphate fractionation, gel filtration on agarose, anion exchange chromatography on DEAE BioGel A, and chromatography on hydroxyapatite. The molar mass of the enzyme in a crude extract was determined by gel filtration on agarose, and found to be 73 000 daltons. The value obtained for the subunit molar mass, by SDS-gel electrophoresis, was 37 000. In its native state the enzyme thus seems to be composed of two subunits and in this aspect resembles its counterparts from animal sources and several bacterial species (Kristjansson, 1977).

Initial velocity studies, carried out at 31 °C with oxaloacetate as the variable substrate and NADH as the changing fixed one, gave a double reciprocal plot with a family of straight intersecting lines in which both the slope and intercept varied (Figure 1). A similar pattern was observed when NADH was used as the variable substrate and oxaloacetate as the changing fixed one (Figure 1). Replots of the slopes and intercepts were linear in both cases (Figure 1, insets). These results are consistent with the operation of a sequential mechanism in which both substrates add to the enzyme prior to the release of products.

The effect of temperature on the kinetic parameters was investigated with experiments conducted at 24.5 °C, 41 °C, 50 °C, 59.5 °C, and 69 °C. A family of intersecting lines with both the slope and intercept changing was obtained at all temperatures except 69 °C, when oxaloacetate was either the variable or changing fixed substrate and NADH the changing fixed, or variable substrate, respectively. Replots of the slopes and intercepts were also linear. The data was thus consistent with the maintenance of the sequential mechanism throughout the temperature range 24.5–59.5 °C. The kinetic constants

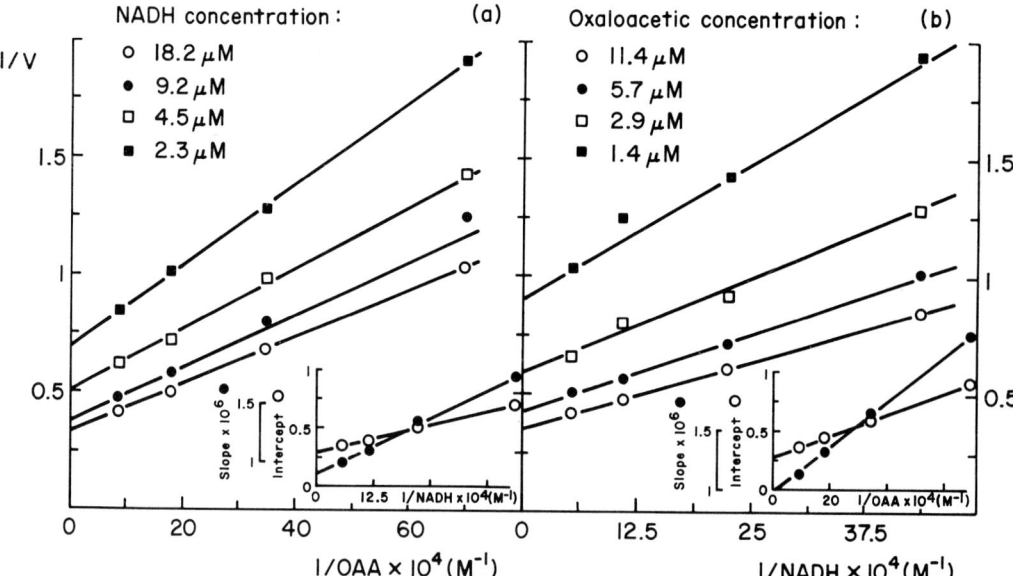

Fig. 1. Initial velocity pattern for MDH at 31 °C. Reciprocal velocities are plotted as a function of OAA (a), or NADH (b), as the varied substrate, with NADH (a), or OAA (b), as the changing fixed one. V is expressed in A_{340} units min^{-1} ml^{-1}. For experimental conditions, see legend with Table I. The insets represent secondary plots of intercepts and slopes from the related primary plots. (a) NADH concentration: -o- 18.2 μM, -•- 9.2 μM, -□- 4.5 μM, -■- 2.3 μM. (b) Oxaloacetic acid concentration: -o- 11.4 μM, -•- 5.7 μM, -□- 2.9 μM, -■- 1.4 μM.

TABLE I
Kinetic parameters for *B. caldolyticus* MDH at various temperatures

Temperature (°C)	24.5	31	41	50	59.5	69
$K_{NADH} \times 10^6$ (M)	5.2 ± 1.3	3.6 ± 0.4	5.7 ± 0.7	12.8 ± 1.5	25.6 ± 6.2	60 ± 70
$K_{OAA} \times 10^6$ (M)	2.9 ± 0.7	3.3 ± 0.3	8.3 ± 1.3	16.7 ± 1.5	45.4 ± 10	460 ± 180
$K_{iNADH} \times 10^6$ (M)	4.1 ± 2.2	2.1 ± 0.5	4.1 ± 1.5	10.1 ± 2.1	38.9 ± 15.5	410 ± 200
V_{MAX} (μmoles NADH/ min/μg MDH)*	0.17 ± 0.02	0.41 ± 0.02	1.0 ± 0.1	1.8 ± 0.1	3.6 ± 0.3	8.8 ± 1.8

*The values reflect an enzyme preparation with specific activity of 93.
All measurements were carried out in 50 mM HEPES buffer, adjusted to pH 8.1 at the temperature under study. The enzyme concentration was 0.034 μg ml^{-1} in all cases, except at 69 °C, where 0.015 μg ml^{-1} was used. At temperatures higher than 41 °C, the stability of NADH and oxaloacetic acid was monitored enzymically. Both were found to be stable at these temperatures, for a period well exceeding the time of reaction.

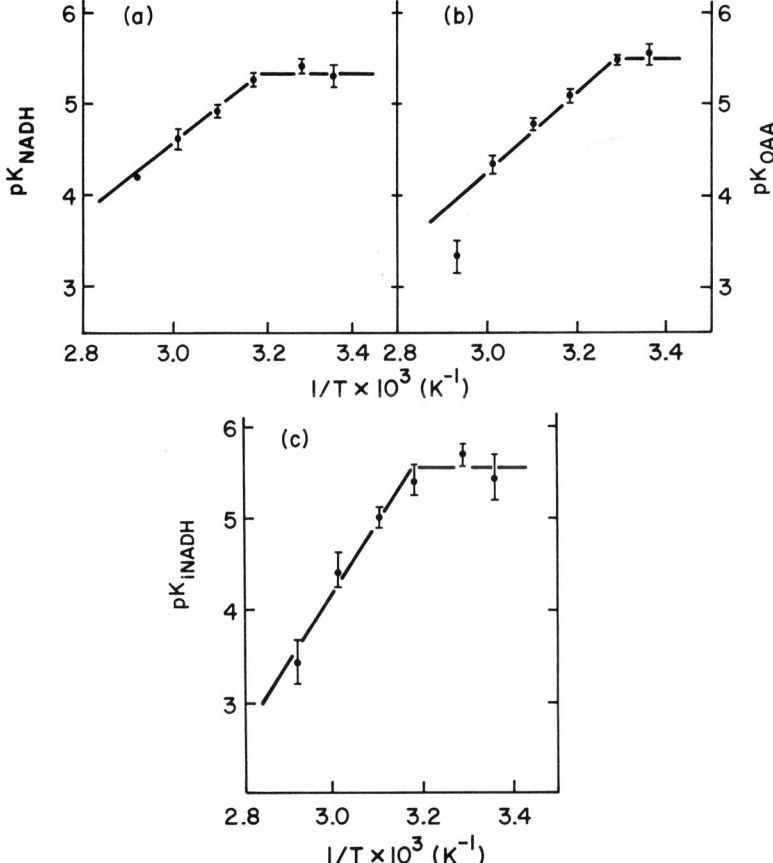

Fig. 2. Van't Hoff plots of the kinetic parameters. The K's are expressed in moles per liter. Bars indicate the standard errors in log K. (a) K_{NADH} (b) K_{OAA} (c) K_{iNADH}.

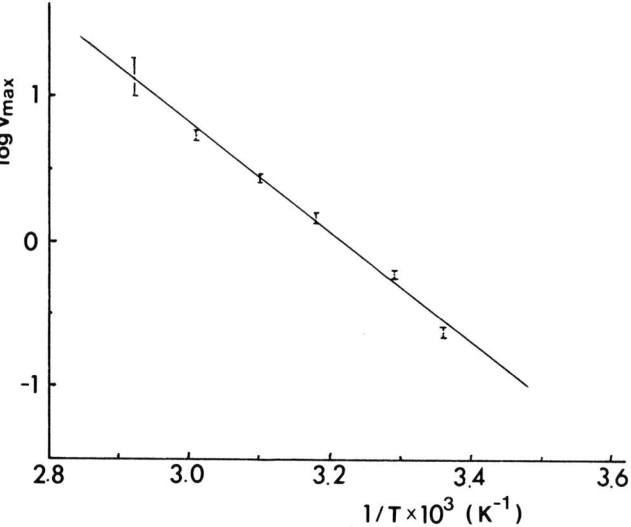

Fig. 3. Arrhenius plot of the *B. caldolyticus* MDH maximal velocity. The velocities are expressed in A_{340} units. Bars indicate the standard errors in log V_{max}.

obtained from these experiments are listed in Table I and are those used by Cleland (1963). NADH is assumed to add first to the enzyme.

Van't Hoff plots of K_{NADH} K_{OAA}, and K_{iNADH}, (Figure 2) all displayed discontinuities at discrete temperatures (T_D's) of 41 °C, 31 °C, and 42 °C, respectively. It is possible that localized or widespread conformational changes in the enzyme at about 42 °C are responsible for the observed discontinuities in the K_{NADH} and K_{iNADH} plots. Stability studies of the enzyme at 41 °C and 51 °C support this view. The data obtained for the entropy of NADH binding to the enzyme showed temperature independence at 41 °C and lower. At higher temperatures the ΔS value decreased with increasing temperature. This could be taken to indicate a conformational change in the enzyme at around 42 °C, which increased the organization of the reaction system. Entropy decreases with enhanced temperature might then be explained by postulating a model where the conformational change of the enzyme involved an additional link(s) between NADH and the enzyme. If the additional connection was of a hydrophobic nature, it would be expected to strengthen upon raising the temperature, thereby increasing the rigidity and decreasing the entropy of the NADH-enzyme complex. Presumably, the access of the NADH-binding site to the substrate and solvent in such a model would be decreased. This model is capable of explaining the initiation of a temperature dependence in the kinetic constants.

The conformational change at T_D, if it exists, was not reflected in discontinuities in maximal velocity as reflected by the Arrhenius plot (Figure 3).

The present data do not allow for the exclusion of other possible explanations for the observed discontinuity. For instance, changes in the rate constants of the reaction, or sudden changes in the pH optimum of the enzyme due to pK changes of functional groups at the active site, might affect the kinetic parameters.

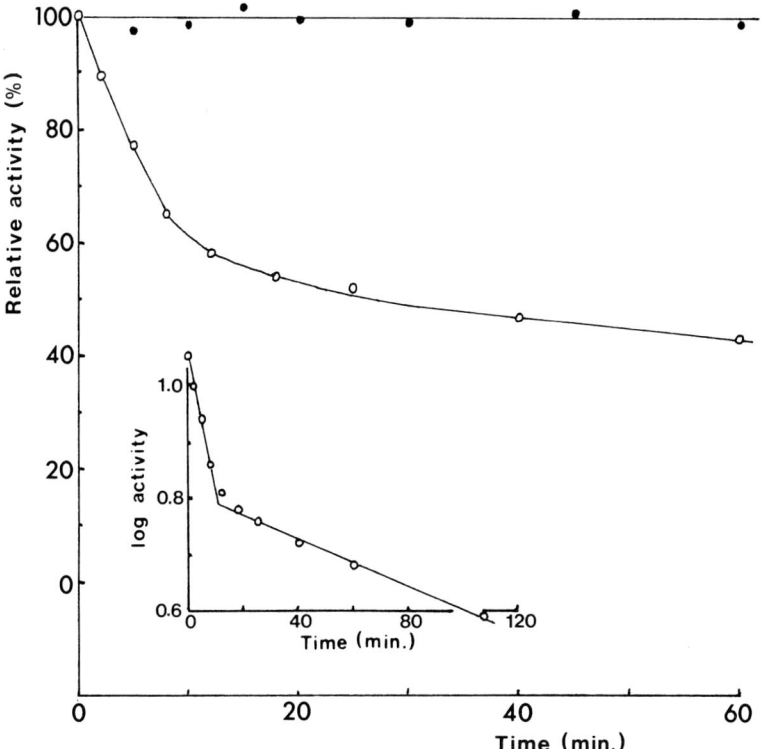

Fig. 4. Heat inactivation of pig heart m-MDH o—o-, and *B. caldolyticus* MDH -•—•-, at 25 °C. The enzyme was preincubated in 50 mM HEPES buffer, adjusted to pH 8.1 at the desired temperature. Assays were conducted at 25 °C (Kristjansson, 1977).

The Arrhenius plot of *B. caldolyticus* malate dehydrogenase shows a straight line in the temperature interval investigated (Figure 3). The activation energy calculated from the plot was 72.2 kJ mole^{-1}. Similar behavior in the Arrhenius plot has been noted for several malate dehydrogenases. The *T. aquaticus* MDH exhibits a straight line Arrhenius plot in the temperature interval 39–70 °C, with an activation energy of 90 kJ mole^{-1} (Biffen and Williams, 1976). Studies on several malate dehydrogenases from animals also reveal a straight line Arrhenius plot, with the activation energy ranging from 43.5–60.6 kJ mole^{-1} (Olle and Olsen, 1975). The *B. caldolyticus* enzyme thus has an activation energy somewhat higher than that observed for the mesophilic enzymes, although less than the E_{act} of a MDH from another extreme thermophile, *T. aquaticus*. The higher activation energy observed for the thermophilic enzymes does imply that these enzymes are not operating as efficiently as their mesophilic counterparts. The loss in efficiency may be compensated for by the higher reaction rates obtained at increased temperatures.

At 69 °C the double reciprocal plot of the kinetic data, where NADH was the variable substrate and oxaloacetate the changing fixed one, did not fit easily to a straight line at cosubstrate concentrations around the expected K_{NADH}. An upward curvature was

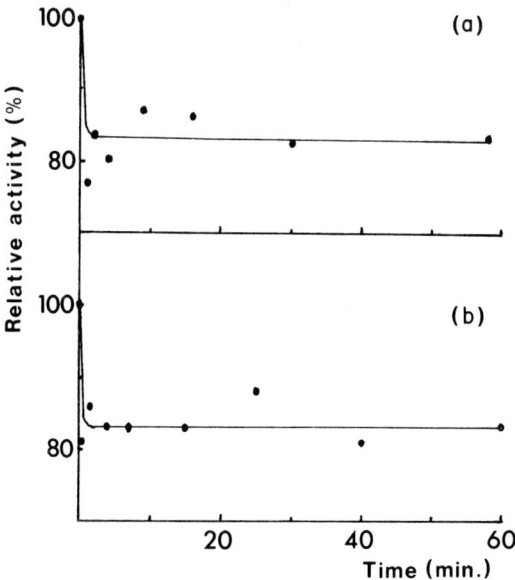

Fig. 5. Heat inactivation of *B. caldolyticus* MDH at 51 °C (a), and 59 °C (b). For experimental conditions, see legend with Figure 4.

obtained at all oxaloacetate concentrations tested. Such data is characteristic of systems showing positive cooperativity and may imply allosteric properties of the enzyme. Also, the appearance of nonlinear kinetics have been explained in terms of a change from an ordered sequence mechanism to a mechanism in which more than one pathway exists for a substrate binding to the enzyme molecule (Sweeney and Fisher, 1968). Thus the possibility for a different reaction mechanism at higher temperatures exists for *B. caldolyticus* malate dehydrogenase. Growth of the bacteria at high temperatures may be related to these observations.

The stability of the *B. caldolyticus* enzyme was examined at a series of temperatures and compared to the stability of pig heart m-MDH at 25 °C. As shown in Figure 4, the pig heart m-MDH activity decreased rapidly with time, under the conditions utilized, where as the *B. caldolyticus* MDH-activity did not change over a time period of one hour. Plotting the log of the pig heart MDH activity as a function of time resulted in the appearance of two lines of different slopes (Figure 4, inset). This suggests a biphasic inactivation of the enzyme.

At 31 °C and 40 °C the *B. caldolyticus* enzyme was unaffected over the 60 min period investigated. As shown in Figure 5, the MDH activity at 51 °C decreased to about 84% of the initial activity before heating, within one minute from the start of the incubation. Thereafter, the activity remained at about the same level during the 60 min time period. At 59 °C, the enzyme behaved in a similar manner, with about 83–84% of the initial activity remaining after 30 s of incubation (Figure 5). At 69 °C, close to the optimum growth temperature of *B. caldolyticus*, the enzyme rapidly lost activity. After 30 s of

Model I

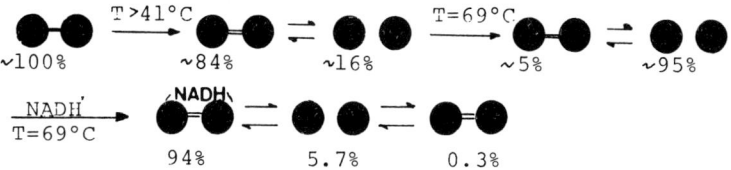

Unconnected dots represent monomers, connected dots represent dimers

Model II

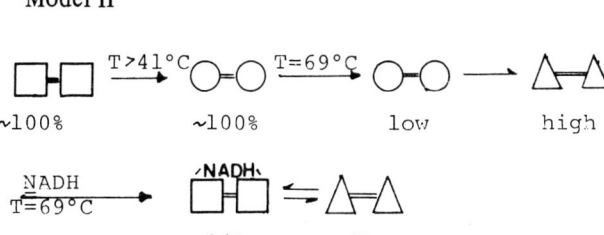

The different symbols refer to enzyme conformations of different activity and stability, as described in the text

Fig. 6. Possible explanations of the stability profiles obtained for *B. caldolyticus* MDH at various temperatures. The percentage values refer to the relative amount of each enzyme species present in reaction mixture.

incubation the activity was about 61% of the initial activity, and was completely lost following 10 min of incubation. The reduction in activity over time had the features of a biphasic inactivation. Such inactivation was previously demonstrated for the pig heart m-MDH at 25 °C. The results at 51 °C and 59 °C also suggest a biphasic inactivation of the enzyme.

In the presence of 5 mM NADH at 69 °C the stability of the enzyme was found to be considerable. It lost about 7% of its initial activity after 30 s of incubation. No further loss of activity was observed following 60 min of incubation. The plot thus resembled the inactivation obtained at 51 °C and 59 °C (Figure 5).

Two different models are postulated for an explanation of the stability profiles compiled (Figure 6). Model I assumes that the same conformation of the enzyme exists at all temperatures investigated. Below 41 °C, the enzyme is thought to exist almost totally in a dimeric state, wheras above 41 °C dissociation into inactive subunits occurs. Based on the stability profiles obtained at 51 and 59 °C, the enzyme at equilibrium is thought to be about 84% in the dimeric state and 16% in the monomeric state. At 69 °C, according to this model, rapid dissociation of the subunits occurs until an equilibrium situation, 5% in dimers and 95% in inactive monomers, is reached. A relatively slow inactivation of the remaining dimers also occurs. At 69 °C in the presence of NADH, the association of the monomers is thought to be favored, resulting in about 94% of the enzyme in a dimeric state. Model I explains the stability profiles reasonably well. A curious feature is that at 51 °C and 59 °C the monomer–dimer equilibrium situation should be identical. Such an occurrence is possible only if strengthening of the hydrophobic interaction compensates for the weakening of other forces as the temperature increases to 59 °C.

Another model to explain the observed profiles is presented in Figure 6. Model II assumes that conformational differences in the enzyme occur at characteristic temperatures. This is consistent with the explanation proposed for the kinetic data. Above 41 °C, according to this view, the enzyme's conformation is changed to a less active but more stable form. At 69 °C, in the absence of large concentrations of NADH, the new conformation changes rapidly to still another conformation, which deactivates slowly. In the presence of NADH, the original conformation is thought to be maintained in an equilibrium with conformation 3. The equilibrium situation would be about 94% in conformation 1 and 6% in conformation 3. Model II thus is consistent with the observed data.

Acknowledgement

The authors are indebted to Dr R. D. MacElroy and Dr L. I. Hochstein NASA/AMES for many fruitful conversations and constant encouragement throughout this work.

References

Biffen, J. H. F. and Williams, R. A. S.: 1976, *Experientia. Suppl.* **26**, 67.
Cleland, W. W.: 1963, *Biochem. Biophys. Acta* **67**, 104.
Kristjansson, H.: 1977, M.S. thesis, University of Maryland.
Olle, S. and Olsen, R.: 1975, *Comp. Biochem. Physiol.* **51B**, 5.
Sweeny, J. R. and Fisher, J. R.: 1968, *Biochem.* **7**, 561.

THE GAS VACUOLE: AN EARLY ORGANELLE OF PROKARYOTE MOTILITY?

JAMES T. STALEY
Department of Microbiology and Immunology, University of Washington, Seattle, Wash. 98195, U.S.A.

Abstract. Several lines of evidence suggest that the gas vesicle may have been an early organelle of prokaryote motility. First, it is found in bacteria that are thought to be representatives of primitive groups. Second, it is a simple structure, and the structure alone imparts the function of motility. Thirdly, it is widely distributed amongst prokaryotes, having been found in the purple and green sulfur photosynthetic bacteria, cyanobacteria, methanogenic bacteria, obligate and facultative anaerobic heterotrophic bacteria, as well as aerobic heterotrophic bacteria that divide by budding and binary transverse fission. Recent evidence suggests that in some bacteria the genes for gas vesicle synthesis occur on plasmids. Thus, the wide distribution of this characteristic could be due to recent evolution and rapid dispersal, though early evolution is not precluded. Though the gas vesicle structure itself appears to be highly conserved among the various groups of bacteria, it seems doubtful that the regulatory mechanism to control its synthesis could be the same for the diverse gas vacuolate bacterial groups.

Tactic responses in microorganisms require motility. Motile heterotrophic bacteria use positive chemotaxis to locate high concentrations of organic substances required for growth. Motile phototrophic bacteria use phototactic mechanisms to locate light needed for photosynthesis. Negative tactic responses also occur in which organisms leave an area where toxic chemicals or high light intensities are inhibitory to growth. Thus, the evolution of motility and its regulation would have represented a significant evolutionary event for organisms. Although there is no direct fossil evidence the first motile organisms probably appeared about 3 billion years ago when anaerobic prokaryotic organisms were the predominant inhabitants of earth.

Bacteria and cyanobacteria exhibit several types of motility. Many unicellular bacteria have flagella by which they propel themselves in aqueous environments. Spirochaetes have an axial filament, consisting of several fibrils analogous to flagella, located inside the outer envelope of the organism (Holt, 1978). These periplasmic "flagella" enable them to move through liquids, especially those of high viscosity (Kaiser and Doetsch, 1975). Certain bacteria and cyanobacteria are able to move by gliding motility while in contact with solid surfaces. Twitching motility is less well understood, but thought to be due to a special type of pilus located at one pole of the organisms (Henrichsen and Blom, 1975). In each of the aforementioned processes the organism actively propels itself by the movement of specialized structures, located in or on the organism. Recent evidence suggests that some gliding organisms have structures located on the external cell envelope which rotate and cause movement while in contact with surfaces in the environment (Jack Pate, personal communication).

In addition to these motility organelles some prokaryotes have gas vacuoles, conferring buoyancy on an organism, permitting it to rise or fall in a water column. Gas vacuole motility is not due to movement of the organelles but to a passive process in which motility is the response caused by the synthesis or dissimilation of an organelle.

Though gas vacuoles were first reported in the late nineteenth century, they were largely ignored by microbiologists prior to electron microscope studies. Bowen and Jensen (1965a, b) 'rediscovered' the structure when they examined the fine structure of several cyanobacteria. Since then our knowledge of gas vacuoles has increased appreciably. (see Walsby, 1972; Walsby, 1978, for reviews).

When observed by phase light microscopy the gas vacuole in cells appears as a bright area with an irregular outline. However, the electron microscope shows the gas vacuole to be comprised of numerous small structures called gas vesicles (Figure 1). The overall appearance is the same in all gas vacuolate organisms examined: a cylindrical midpiece approximately 40–100 nm wide that terminates with a conical section at each end (Cohen–Bazire *et al.*, 1969). The organelle is unique, consisting of a thin protein membrane surrounding a space. The membrane is impermeable to water, but freely permits the passage of gases. Since inert gases such as argon readily pass through it, the vacuoles apparently do not selectively store gases (Walsby, 1969).

The gas vacuole is thought to regulate buoyancy. The gas vesicles of the vacuole reduce cell density, and if the decrease in density is sufficiently great, the organism will ascend in a water column. Conversely, by reducing the volume occupied by the vesicles through cell growth or vesicle breakdown, the cell's density is increased thereby enabling it to descend.

Several lines of evidence suggest that gas vacuoles were early organelles of motility. The gas vacuole is a simple structure whose function is imparted by the structure alone. In contrast, organisms which move by flagellar, or other types of active motility require, in addition to the necessary structures, an apparatus for converting chemical energy to mechanical energy, a process which would, *a priori*, appear to be much more complex and hence, more recently evolved.

Furthermore, the gas vacuole has been found in organisms derived from primitive organisms, such as anaerobic bacteria in the genera *Clostridium* and *Methanosarcina. Clostridium* spp. are fermentative heterotrophs that do not synthesize heme-containing proteins such as catalase and cytochromes and, for these reasons, have been regarded as possible early organisms (Broda, 1975; Margulis, 1970). Recently, gas vacuolate strains of anaerobic spore-forming bacteria (i.e. *Clostridium* spp.) have been reported from soil habitats (Krasilnikov *et al.*, 1971). Likewise, some of the so-called archaebacteria contain gas vacuoles, for example, the methane producing *Methanosarcina barkeri* which grows anaerobically on acetate, methanol, or carbon dioxide and hydrogen and produces methane gas (Zhilina, 1971; Mah *et al.*, 1977). Methanogens have been suggested as likely candidates for the earliest microorganisms based on the sequence of nucleotides in their 16S ribosomal RNA (Woese and Fox, 1977). Many strains of the archaebacterial genus *Halobacterium* also produce gas vacuoles; however since these are aerobes they probably appeared after the transition to an oxidizing atmosphere.

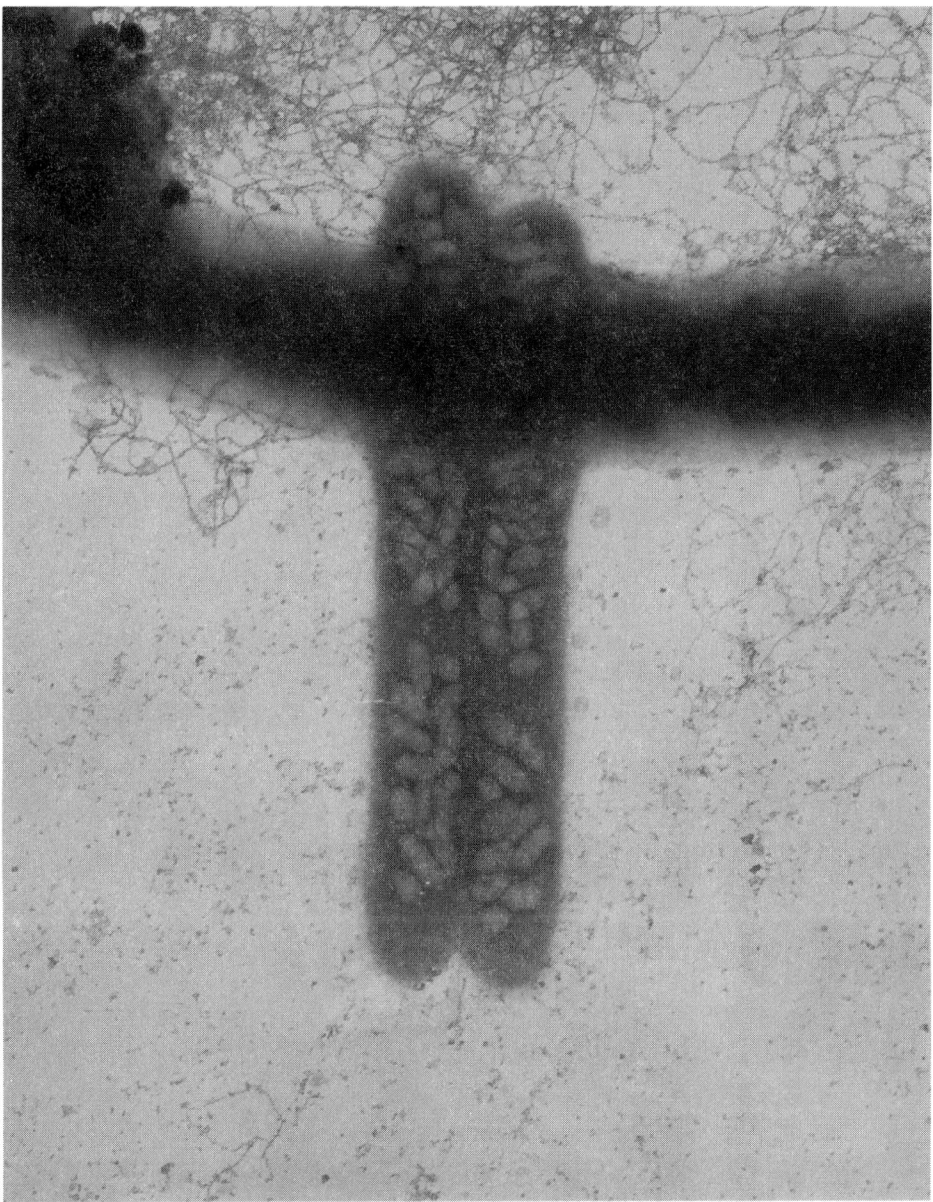

Fig. 1. Two cells of a gas vacuolate bacterium from the hypolimnion of a lake. These cells each contain about 35 gas vesicles of the characteristic shape. Note that some gas vesicles are smaller than others. A filamentous microbe is lying over the two cells obscuring them partly. Electron micrograph of whole cells fixed with Lugol's iodine.

Also consistent with an early evolution of the gas vacuole is its widespread distribution amongst prokaryotes. Not only is the structure found among presumed early anaerobic, heterotrophic and methanogenic bacteria, but it is also widely distributed among photosynthetic bacteria, representatives of which are thought to have evolved about 3 billion years ago (Schwartz and Dayhoff, 1978).

The gas vacuole was probably as important in bringing early photosynthetic organisms close to the surface of aquatic habitats as it is today. Light does not penetrate deeply into water because of its high absorption by water. Organisms requiring sunlight as an energy source would need some mechanism to move into the photic zone, and the gas vacuole may have been the earliest solution to this problem. Although none of the known green sulfur bacteria possess flagella, many species do have gas vacuoles. Included among the gas vacuolate green sulfur bacteria are strains of *Pelodictyon, Clathrochloris* and *Ancalochloris* (Buchanan and Gibson, 1974). Many of the purple sulfur bacteria also presumed to have been early photosynthetic organisms, contain gas vacuoles, for example, representatives of the genera *Lamprocystis, Thiodictyon, Thiopedia* and *Amoebobacter*. The other major group of photosynthetic prokaryotes presumed to have derived from the anaerobic photosynthetic bacteria are the cyanobacteria. Gas vacuolate representatives included in this group are strains of the genera *Anabaena, Spirulina, Nostoc, Oscillatoria, Microcystis, Gloeotrichia, Aphanizomenon, Trichodesmium, Calothrix, Anabaenopsis, Phormidium, Lyngbya, Coelosphaerium*, and *Tolypothrix*.

In addition to *Clostridium*, other anaerobic heterotrophic bacteria have been reported to contain this structure, including the genus *Meniscus* (Irgens, 1977). Others such as *Brachyarcus* have been observed in the anaerobic hypolimnion of stratified lakes (Caldwell and Tiedje, 1975; Clark and Walsby, 1978). Facultative anaerobes have also been isolated, one of which is the prosthecate bacterium *Ancalomicrobium adetum* (Staley, 1968). Another remains unidentified (Van Ert and Staley, 1971b).

Filamentous gas vacuolate organisms in the genera *Pelonema* and *Peloploca* have not been isolated so their requirements for oxygen have not been determined.

The final group of gas vacuolate heterotrophic bacteria is the aerobic heterotrophic bacteria. The multiple appendaged prosthecate bacterium, *Prosthecomicrobium pneumaticum* (Staley, 1958) and the ring-forming *Microcyclus aquaticus* (Van Ert and Staley, 1971a) are representative of this group.

The examples mentioned amply illustrate the widespread occurrence of the gas vacuole among prokaryotic organisms. It is more widespread than any other organelle of motility, including the flagellum. Gas vacuoles are found in gram positive and in gram negative bacteria as well as the methanogens, in photoautotrophs and heterotrophs, aerobes as well as anaerobes, budding bacteria as well as bacteria that divide by binary transverse fision. All cyanobacteria and all green sulfur bacteria lack flagella.

An alternative hypothesis to explain the widespread occurence of the gas vacuole among prokaryotes is that genes for gas vacuole synthesis may occur on plasmids which are readily transferred among organisms. The structure may have evolved recently and may have been dispersed widely by plasmid transfer, as apparently is the case of *Halobac-*

terium spp. (Simon, 1978; Weidinger *et al.*, 1979). A recent origin and plasmid transmission of vacuoles implies that structurally and perhaps functionally the gas vacuole would be quite uniform from one species to the next. Although there is some variation in the size of gas vacuoles, their shape is uniform. Amino acid analyses conducted on the vacuolar membrane protein indicate similarities, such as the absence of sulfur-containing amino acids and high concentrations of hydrophobic amino acids while showing differences in the molar composition of the amino acids. Serological analyses using antiserum from purified gas vesicles of *Microcyclus aquaticus* indicated that they were sufficiently similar to cross-react with gas vesicles or cell lysates of all organisms tested including *Halobacterium, Prosthecomicrobium pneumaticum*, and the blue green bacteria *Nostoc muscorum* and *Anabaena flos-aquae* (Konopka et al., 1977).

These similarities in structure and composition may not be due to plasmid dispersal but rather conservation of the structure in order not to lose its function. Too great a variation in structure may lead to loss of hydrophobic characteristics and function which has been selected against.

The largest gap in our knowledge concerns the regulation of formation of this structure (Konopka, 1977). The plasmid dispersal hypothesis would appear to suffer most from what is known in this respect. The control mechanisms regulating the synthesis of this structure in the photosynthetic, anaerobic, multicellular *Pelodictyon clathratiforme* probably differ from those in the heterotrophic, aerobic, unicellular *Microcyclus aquaticus*, an argument against the plasmid dispersal hypothesis. A considerable degree of modification would have been needed if a plasmid containing this structure were transferred from one of these gas vacuolate representatives to the other. Over the course of a billion years, the evolutionary process could account for the gas vacuole variations as new groups evolved.

References

Bowen, C. C. and Jensen, T. E.: 1965a, *Science* **147**, 1460.
Bowen, C. C. and Jensen, T. E.: 1965b, *Am. J. Bot.* **52**, 641.
Broda, E.: 1975, *The Evolution of the Bioenergetic Process*, Pergammon Press, Oxford.
Buchanan, R. E. and Gibbons, N. E. (eds.): 1974; *Bergey's Manual of Determinative Bacteriology*, Williams and Wilkins, Baltimore, Maryland.
Caldwell, D. E. and Tiedje, J. M.: 1975, *Can. J. Microbiol.* **21**, 362.
Clark, A. E., and Walsby, A. E.: 1978, *Archives Microbiol.* **118**, 229.
Cohen-Bazire, G., Kunisawa, R., and Pfennig, N.: 1969, *J. Bacteriol.* **100**, 1049.
Henrichsen, J. and Blom, J.: 1975, *Acta. Path. Microbiol. Scand. Sect.* **B83**, 161.
Holt, S. C.: 1978, *Microbiol. Rev.* **42**, 114.
Irgens, R. L.: 1977, *Int. J. Syst. Bacteriol.* **27**, 38.
Kaiser, G. E. and Doetsch, R. W.: 1975, *Nature (London)* **255**, 656.
Konopka, A. E., Lara, J. C., and Staley, J. T.: 1977, *Arch. Microbiol.* **112**, 133.
Konopka, A. E.: 1977, *Can. J. Microbiol.* **23**, 363.
Krasilnikov, N. A., Duda, V. I., and Pivovarov, G. E.: 1971, *Microbiology* (Engl. trans.) **40**, 592.
Mah, R. A., Ward, D. M., Baresi, L., and Glass, T. L.: 1977, *Ann. Rev. Microbiol.* **31**, 309.
Margulis, L.: 1970, *Origin of Eukaryotic Cells*, Yale University Press, New Haven.
Schwartz, R. M. and Dayhoff, M. O.: 1978, *Science* **199**, 395.
Simon, R. D.: 1978, *Nature* **273**, 314.

Staley, J. T.: 1968, *J. Bacteriol.* **95**, 1921.
Van Ert, M. and Staley, J. T.: 1971a, *J. Bacteriol.* **108**, 236.
Van Ert, M. and Staley, J. T.: 1971b, *Arch. Microbiol.* **80**, 70.
Walsby, A. E.: 1969, *Proc. Roy. Soc. London* **B173**, 235.
Walsby, A. E.: 1972, *Bacteriol. Rev.* **36**, 1.
Walsby, A. E.: 1978, in *Symposium of the Society for General Microbiology* **28**, 327.
Weidinger, G., Klotz, G., and Goebel, W.: 1979, *Plasmid* **2**, 377.
Woese, C. R. and Fox, G. E.: 1977, *Proc. Natl. Acad. Sci.* **74**, 5088.
Zhilina, T. N.: 1971, *Microbiology* **40**, 587.

PHYSICAL CHEMISTRY AND EVOLUTION OF SALT TOLERANCE IN HALOBACTERIA

JANOS K. LANYI

NASA-Ames Research Center, Moffett Field, Calif., U.S.A.

Abstract. The cellular constituents of extremely halophilic bacteria not only tolerate high salt concentration, but in many cases require it for optical functioning. The characteristics affected by salt include enzyme activity, stability, allosteric regulation, conformation and subunit association. The salt effects are of two major kinds: electrostatic shielding of negative charges by cations at low salt concentration, and hydrophobic stabilization by salting-out type salts at high salt concentration. The composition of halobacterial proteins shows an excess of acidic amino acids and a deficiency of non-polar amino acids, which accounts for these effects. Since the cohesive forces are weaker and the repulsing forces are stronger in these proteins, preventing aggregation in salt, these structures are no longer suited for functioning in the absence of high salt concentrations. Unlike these nonspecific effects, ribosomes in halobacteria show marked preference for potassium over sodium ions. To ensure the proper intracellular ionic composition, powerful ion transport systems have evolved in the halobacteria, resulting in the extrusion of sodium ions and their replacement by potassium. It is likely that such membrane transport system for ionic movements is a necessary requisite for salt tolerance.

Interest in the halobacteria occurs for at least three reasons. They belong to the group, 'archaebacteria', together with the methanogens and the thermoacidophiles (Woese *et al.*, 1978), whose members are supposedly of ancient lineage. They contain a unique light-reactive membrane protein, bacteriorhodopsin, which serves as a source of energy for the cells (Lanyi, 1978; Stoeckemius *et al.*, 1979). They are limited to an extreme habitat: these organisms grow and survive only at high concentrations of salt, typically above 3 molar NaCl.

Given the fact that the concentrations of salt that these bacteria tolerate is between 3 and 5 molar, it is an outstanding and most important observation that the internal salt concentration is also very high (Larsen, 1967; Bayley and Morton, 1978; Kushner, 1978). In fact, the cells appear to be in osmotic equilibrium with the outside, and for the halobacteria salt itself serves as the internal osmoticum. The way in which the intracellular components of these cells respond to salt is therefore an important part of the larger question of cellular salt tolerance.

The enzymes of halobacteria show a variety of responses to salt. For example the enzyme activity of aspartate transcarbamylase from *Halobacterium cutirubrum* is absolutely dependent on salt (Liebl *et al.*, 1969). There is virtually no activity in the absence of NaCl or KCl, and maximal activities are observed at about 4 molar salt, corresponding fairly closely to the internal salt concentration in these cells.

Another kind of response is exhibited by isocitric dehydrogenase from the same organism. Here, the enzyme activity does not require salt, but is able to survive the normally deleterious effects of salt, and some enhancement of activity is observed at intermediate salt concentrations (Aitken et al., 1970; Brown, 1976). Some specificity for salt is observed for this enzyme: potassium chloride is more effective at higher concentrations than sodium chloride, although the differences are not dramatic.

The third kind of response to salt is that of fatty acid synthetase. Halobacteria do not contain a large quantity of fatty acids, since their lipids are of the ether type (Kales, 1972), but they do contain the synthetase. This enzyme is inhibited by salt (Pugh et al., 1971), but less so than, nonhalophilic enzymes are. Although different kinds of salt dependent behavior are exhibited by halobacterial enzymes, they always show considerable enzyme activity even at very high concentrations of salt (for a review, see Lanyi, 1974).

Halophilic NADH-menadione oxidoreductase has been extensively studied (Lanyi, 1969; Lanyi and Stevenson, 1970; Lanyi, 1972). Enzyme activity, maximal between 2 and 3 molar NaCl, in this system is strictly dependent on salt. The stability of the enzyme is also dependent on salt. When the enzyme was exposed to salt solutions of different concentrations, with salt added at different times to raise its concentration above 3 molar, the time dependent decay of the enzyme activity could be described with a single exponential, and therefore with a first-order rate constant. The rate constant for the spontaneous inactivation is greatest at low salt concentrations, varying exactly as the inverse of enzyme activity (Lanyi, 1969). Thus, salt enhances the stability and activity of this enzyme.

Other enzyme properties are also salt dependent. The salt dependence and the regulation of the allosteric threonine deaminase from *H. cutirubrum* by ADP, for example, are intimately related (Lieberman and Lanyi, 1972). Other investigators have found that the subunit association of some of these halophilic enzymes is dependent on salt (Norberg et al., 1973; Aitken and Brown, 1972; Mevarech and Neumann, 1977). Finally, it turns out that the structural integrity of larger cellular structures, such as membranes (Lanyi, 1975) and ribosomes (Bayley and Kushner, 1964) requires salt also.

As so much of what characterizes macromolecules from halobacteria depends on salt, the underlying reasons for these effects are fertile ground for exploration. The explanation should include the evolutionary modifications made in these structures, in physical chemical terms, and how these features will cause them to give the observed effects of salt.

Unfortunately, satisfactory explanations have not yet been given. In fact, only generalization, or what is worse, speculations over generalizations have been offered. One reason is that until recently a purified enzyme from halobacteria has not been available in large enough quantities for physical chemical studies. However, some clues were obtained about the nature of the salt dependence by treating the enzymes under different conditions and following the consequences for catalytic activity.

The kinetic properties of halobacterial NADH-menadione oxidoreductase were determined in different salts. Varying the cation made little difference, but the nature of the anion was of great importance. For example, NaCl was distinctly preferred for supporting enzyme activity over $NaNO_3$ (Lanyi and Stevenson, 1970). The order in which the salts supported enzyme activity fell into the lyotropic series, suggesting that hydrophobic interactions were involved in the structure of the enzyme. Such bonds arise from the tendency of non-polar molecules to organize water in their vicinity. When an organic substance is dissolved in water, the solvent will be more highly structured around the molecules, and this increase in organization is entropically unfavorable. When proteins are denatured, groups which were previously in the inside become exposed to the exterior and to water. This process is somewhat analogous to the dissolution of organic molecules, since water will be organized around these parts of the protein. Some of the stability of proteins is, in fact, derived from the unfavourable entropy change during the unfolding of the polypeptide chains and the exposure of buried groups.

Those salts that decrease the solubility of organic compounds in water, i.e. the 'salting-out type' salts, are expected to stabilize hydrophobic bonding in proteins, and the more a protein relies on such bonds for stability the more it will be stabilized. This idea seems to apply to enzymes from halobacteria. We have concluded that above 1 molar concentration, the salt dependence of halophilic proteins is related to strengthening of hydrophobic bonds (Lanyi, 1974).

The effects of salt are somewhat more complicated in the case of the cytochrome oxidase of *H. cutirubrum* (Lieberman and Lanyi, 1971). Here the salt dependence can be resolved into two phases: at low and high concentrations of salt. Low salt concentrations give partial enzyme activity, which does not depend on the nature of the anion or cation. This partial activity is also obtained with magnesium or calcium chloride, at millimolar concentrations. At higher salt concentrations, e.g. above 1 molar, the nature of the anion makes a much greater difference. It appears, therefore, that there are two distinct phenomena. At low salt concentrations, salt seems to be the shielding of charges, depending on the location of charged residues in the protein and their screening by cations, while at high concentrations salt seems to influence the hydrophobic interactions.

The idea that hydrophobic bonds are involved in salt dependence was reinforced by studies of the thermodynamics of enzyme inactivation. As mentioned above, the decay of NADH-menadione oxidoreductase from *H. cutirubrum* at lowered salt concentrations can be characterized by a first-order rate constant. Arrhenius plots of enzyme inactivation consist of graphs of the logarithm of this rate constant vs. the reciprocal absolute temperature. For most enzymes this kind of plot will be linear, with a negative slope, i.e. the enzymes are less stable at higher temperatures. In the case of the halophilic NADH-menadione oxidoreductase (Lanyi and Stevenson, 1970) the plot exhibits a reversal: at lower as well as at higher temperatures there is more rapid inactivation, and the enzyme thus shows an optimum temperature for stability (at about 20 °C). While such behavior is not unknown for enzymes, it is unusual. Cold lability is a property of hydrophobic bonds. At lower temperatures water structure is increased, and thus the tendency for exposure of non-polar amino acid residues, normally buried inside the protein, is enhanced. Cold lability is shown by most halobacterial enzymes and is greater at higher salt concentration (Lanyi and Stevenson, 1970). At lower salt concentrations the enzymes are denatured more rapidly, but show less cold lability, suggesting that the salt dependent increase in stability is specifically sensitive to the lower temperature.

From such considerations we have put together a general conceptual model for the effects of salt on halophilic proteins (Lanyi, 1974). This model does not refer to any single protein: its elements originated from results with different enzymes and represents an illustration of an idea. Surprisingly enough, virtually no subsequently obtained results have seriously contradicted this model, and thus it does seem to have some predictive power.

The essential point of the model is that in salt dependence we are dealing with two kinds of effects. Charge screening by cations at lower salt concentrations will cause some specific folding of the polypeptide chains. This is largely absent without the salt. This structure will generally give partial enzyme activity, and in some cases lower affinity to substrates. If then higher salt concentrations e.g., between 1 and 4 molar, of salting-out type salts are introduced, the enzyme will assume its optimal conformation with maximal activity and stability. The low-salt forms of the enzymes are unstable, and their decay is reflected in the inactivation rate constants. We can shift the equilibrium among the different conformations by adding different amounts of salt, and thereby vary the apparent rate constant of inactivation. If the salt added is the salting-in type, then charge screening, but not hydrophobic stabilization, is obtained. Specific examples, such as specific purified proteins from halobacteria, are necessary to test this model and very few of these are available.

For statistical purposes ribosomal proteins were chosen, because the amino acid compositions were available, and because we did not wish to prejudice ourselves by the selection of specific proteins. Proteins of 50S ribosomes from *Escherichia coli* and *H. cutirubrum* were compared (Lanyi, 1974). It was evident that in the halobacterial proteins the amount of acidic minus basic amino acids is greatly increased relative to the *E. coli* proteins. Halobacterial proteins on the average are quite acidic, while the *E. coli* ribosomal proteins tend to be basic. Individual examples for such acidity have been described before for halobacterial enzymes (Larsen, 1967). Additionally, the halophilic proteins seem to be deficient in hydrophobic amino acids. The observed excess of acidic amino acids and deficiency of hydrophobic amino acids account very well for the model of the salt effects described above, since both of these conditions will result in structural instability, which is overcome by salt effects.

If we consider some of the borderline amino acids which possess intermediate chain length and do not ordinarily participate in hydrophobic bonding, their frequency is greater in the halobacterial proteins relative to the *E. coli* proteins. It seems likely that high concentrations of salting-out type salts will cause these residues to become involved in hydrophobic interactions that could not be formed in water in the absence of salt. A convenient point of reference for these comparisons is provided by the frequencies of glycine and alanine, which do not contribute to the stability of proteins. The frequencies of these amino acids are equivalent in halobacteria and *E. coli* (Lanyi, 1974).

The model for salt dependence is self-consistent, but one might ask what evolutionary pressure could have altered these proteins in such a way. The salt dependence seems to be a liability, rather than an advantage. The answer is that if proteins from non-halophilic organisms are placed in several molar salt, they will undergo several kinds of structural changes, all of which are deleterious to enzyme activity. The polypeptide chains will collapse into a tightly folded structure, and separate structures will aggregate, thereby losing functional properties. This can be overcome by decreasing the cohesive forces in the proteins: by modifying them to lessen the possibility for hydrophobic bonding, and by increasing the number of charged residues to cause electrostatic repulsion. Such modified enzymes should function very well at high salt concentrations, but they will have lost the ability to maintain their structure without the salt. According to this idea, evolutionary adaptation to salt leads to full committment to such an environment, and thus salt tolerance is intimately linked to salt dependence.

However, there is a possibility for making enzymes which will function at both high and low salt concentrations. Such enzymes would contain increased numbers of charged residues and decreased numbers of hydrophobic residues, but would be decreased in size. On thermodynamic grounds carboxyl residues are known to be on the outside of proteins, and hydrophobic groups are on the inside. If the molecular weight of the protein is decreased, the surface to volume ratio is increased, and it is possible to accommodate these changes in composition without penalties. There is some evidence (Fitt, 1978), that some proteins, one known enzyme and many which have not been identified with respect to function, are much smaller than expected when compared with proteins from non-halophilic bacteria.

Everything discussed above with respect to salt made no distinction between potassium and sodium chlorides. However, the halobacteria make a very important distinction between these two salts, as they exclude sodium and accumulate potassium even against very high gradients across the cytoplasmic membrane (Lanyi, 1978). Since the cells expend large amounts of energy in doing this, the difference between Na^+ and K^+ must be critical for intracellular processes. Most halophilic enzymes show only a mild preference for potassium, however (Brown, 1976; Lanyi, 1974), which should not be sufficient to warrant great energy expenditures for replacing sodium with potassium. There is one exception to this, and it is probably a decisive one: ribosomal structure. The ribosomes from *H. cutirubrum* are stable only in the presence of 4 molar potassium chloride, and sodium chloride will not substitute (Bayley and Kushner, 1964). Although the physical chemistry of this cation specificity is not yet understood, it evidently accounts for the requirement of potassium in the cytoplasm. It appears, therefore, that the mechanism for accumulating potassium and excluding sodium is an important aspect of salt tolerance.

Much of the information on cation transport in the halobacteria has been obtained from studies with cell envelope membrane vesicle preparations (MacDonald and Lanyi, 1975; Lanyi *et al.*, 1976). The energy input into these vesicles may be supplied with illumination of bacteriorhodopsin, in a retinal-protein pigment found in these membranes (Lanyi, 1978; Stoeckenius *et al.*, 1979), which initiates the circulation of protons between the inside and the outside of the vesicles. The protons extruded by bacteriorhodopsin will recirculate through a specific membrane component coupling the proton influx to the extrusion of sodium ions. The sodium ions will also recirculate and thereby drive the transport of amino acids. The coupling of the sodium gradient to the transport of the amino acids occurs in such a way as to suggest that the sodium ions and the amino acids are cotransported through specific transport carriers (Lanyi et al., 1976; MacDonald et al., 1977). While sodium is removed very efficiently from the vesicle (and cell) interior, potassium ions are taken up, apparently in response to the electrical potential difference across the membranes (Lanyi, 1978). Potassium uptake is much more effective in intact cells than in membrane preparations (Lanyi, 1978), indicating that an essential component of the potassium transport system is lost during the isolation of the membranes. To compensate for the slower uptake of potassium, the membrane vesicles lose chloride during illumination (Lanyi et al., 1979). The resulting energization will cause the vesicles to either replace sodium with potassium when potassium influx is high, or lose both sodium and chloride leading the water loss and collapse. The latter process offers the means for volume regulation.

In principle, sodium transport has essentially two reasons for being; the first is to remove sodium from the cell interior, which seems to be a requirement for optimal cytoplasmic function as described above, and the second is to participate in the energetics of membrane processes, both by driving the active transport of metabolites and by influencing the gradient of protons (Lanyi and MacDonald, 1976). An important question is which of these mechanisms came first. From an evolutionary point of view, is sodium transport primarily a mechanism for maintaining the proper cationic environment of the cell interior, or is it primarily a mechanism for better management of the energy resources of the cells? Skulachev recently suggested (Skulachev, 1978) that from an energetic point of view sodium gradients have advantages over proton gradients. He argues that sodium gradients have much higher capacities than proton gradients and thus represent longer term reservoirs of energy, and also that larger sodium gradients can be produced than proton gradients because when the pH gradient is large the cytoplasmic pH rises to unphysiological levels.

These arguments appear to carry weight in the case of the halobacteria, but may not apply to non-halophilic bacteria. In the latter organisms the capacities of sodium and proton gradients are about equal (Schulz et al., 1963; Harold and Papinean, 1972), and thus, the pH buffering capacity of the cell interior is not very different from the capacity of the system for sodium ions. The second point is not very compelling as the internal pH in various kinds of cells is, in fact, constant (Ramos et al., 1976; Padan et al., 1976; Guffanti et al., 1978). Bacterial cells near neutrality do not maintain a pH difference across the cytoplasmic membrane, but rely on membrane potential for driving energy-requiring functions. Thus, energy is not stored in a concentration difference for protons, and the problem of high cytoplasmic pH does not arise.

The evolutionary reasons for sodium transport are still unresolved, but the evidence favors the idea that it arose primarily to maintain low cytoplasmic sodium levels, favorable for some structures or reactions. If it is a mechanism for salt tolerance, the energetic costs are certainly high, particularly for the halobacteria.

References

Aitken, D. M. and Brown, A. D.: 1972, *Biochem. J.* **130**. 645.
Aitken, D. M., Wicken, A. J., and Brown, A. D.: 1970, *Biochem. J.* **116**, 125.
Brown, A. D.: 1976, *Bacteriol. Rev.* **40**, 803.
Bayley, S. T. and Kushner, D. J.: 1964, *J. Mol. Biol.* **9**, 654.
Bayley, S. T. and Morton, R. A.: 1978, *CRC Crit. Rev. Microbiol.* **6**, 151.
Fitt, P. S.: 1978 in S. R. Caplan and M. Ginzburg (eds.), *Halophilic Microorganisms*, Elsevier, Amsterdam, pp. 379–396.
Guffanti, A. A., Susman, P., Blanco, R., and Krulwich, T. A.: 1978, *J. Biol. Chem.* **253**, 708.
Harold, F. M. and Papineau, D.: 1972, *J. Memb. Biol.* **8**, 45.
Kates, M.: 1972, in F. Snyder (ed.), *Ether Lipids. Chemistry and Biology*, Academic Press, New York, pp. 351–398.
Kushner, D. J.: 1978, in D. J. Kushner (ed.), *Microbial Life in Extreme Environments*, Academic Press, London, pp. 317–368.
Lanyi, J. K.: 1969, *J. Biol. Chem.* **244**, 4168.
Lanyi, J. K.: 1972, *J. Biol. Chem.* **247**, 3001.
Lanyi, J. K.: 1974, *Bacteriol Rev.* **38**, 272.
Lanyi, J. K.: 1975, in M. R. Heinrich (ed.), *Extreme Environments: Mechanisms of Microbial Adaptation*, Academic Press, New York, pp. 295–303.
Lanyi, J. K.: 1978, *Microbiol. Rev.* **42**, 682.
Lanyi, J. K. and MacDonald, R. E.: 1976, *Biochemistry* **15**, 4608.
Lanyi, J. K. and Stevenson, J.: 1970, *J. Biol. Chem.* **245**, 4074.
Lanyi, J. K., Yearwood-Drayton, V., and MacDonald, R. E.: 1976, *Biochemistry* **15**, 1595.
Lanyi, J. K., Renthal, R., and MacDonald, R. E.: 1976, *Biochemistry* **15** 1603.
Lanyi, J. K., Helgerson, S. L., and Silverman, M. P.: 1979, *Arch. Biochem. Biophys.* **193**, 329.
Larsen, H.: 1967, *Adv. Microbial Physiol.* **1**, 97.
Liebl, V., Kaplan, J. G., and Kushner, D. J.: 1969, *Can. J. Biochem.* **47**, 1095.
Lieberman, M. M. and Lanyi, J. K.: 1971, *Biochim. Biophys. Acta* **245**, 21.
Lieberman, M. M. and Lanyi, J. K.: 1972, *Biochemistry* **11**, 211.
MacDonald, R. E. and Lanyi, J. K.: 1975, *Biochemistry* **14**, 2882.
MacDonald, R. E., Greene, R. V., and Lanyi, J. K.: 1977, *Biochemistry* **16**, 3227.
Mevarech, M. and Neumann, E.: 1977, *Biochemistry* **16**, 3786.
Norberg, P., Kaplan, J. G., and Kushner, D. J.: 1973, *J. Bacteriol.* **113**, 680.
Padan, E., Zilberstein, D., and Rottenberg, H.: 1976, *Eur. J. Biochem.* **63**, 533.
Pugh, E. L., Wassef, M. K., and Kates, M.: 1971, *Can. J. Biochem.* **49**, 953.
Ramos, S. Schuldiner, S., and Kaback, H. R.: 1976, *Proc. Nat. Acad. Sci. U.S.A* **73**, 1892.
Schultz, S. G., Epstein, W., and Solomon, A. K.: 1963, *J. Gen. Physiol.* **47**, 329.
Skulachev, V. P.: 1978, *FEBS Lett* **87**, 171.
Stoeckenius, W., Lozier, R. H., and Bogomolni, R. A.: 1979, *BBA Bioenergetics* **505**, 215.
Woese, C. K., Magrum, L. J., and Fox, G. E.: 1978, *J. Mol. Evol.* **11**, 245.

HALOPHILY AND HALOTOLERANCE IN CYANOPHYTES

STJEPKO GOLUBIC

Boston University Boston, Mass. 02215, U.S.A.

(Received 19 February, 1980)

Abstract. The survival, growth and distribution of organisms in hypersaline environments is discussed using cyanophytes (cyanobacteria) as examples. The distinction between halophilic (Na^+-requiring) and halotolerant organisms is not adequate to describe the entire spectrum of adaptations to salt. The classical division into stenohaline (narrow) and euryhaline (wide) adaptational types, with optima identified as oligo-, meso- and polyhaline, better reflects both organismal adaptations and the environmental conditions to which these are adjusted and is therefore recommended as a conceptual model.

Two independent properties of organisms are growth and survival. Organisms requiring narrow ranges of salt concentration are considered specialists and are restricted to environments with relatively constant salinities at any particular concentration. Organisms which tolerate wide ranges of fluctuation in salinity are considered generalists. The existence of separate and distinct microbial assemblages in these two types of environments is demonstrated in marine intertidal zones and seasonal salt works, representative of fluctuating salinity, and in the open ocean. The hypersaline ponds of Yallahs, Jamaica, and Solar Lake, Sinai represent different but relatively constant salinities. It is concluded that cyanophytes speciate along the salinity gradient, and that separate halophilic taxa occupy environments with relatively constant salinities.

1. Introduction

Many aquatic environments become periodically reduced in volume by evaporation, passing through stages of increasing ionic concentrations and sequential mineral precipitation. Such conditions are common in arid regions of the world, and, judging from the abundance of evaporitic sediments in the fossil record, have also been common in the past. Evaporation posed difficulties to the survival of aquatic organisms, which responded by evolving a variety of mechanisms to cope with high salinities; this distinguishes them from freshwater organisms. To express this distinction such terms as halotolerant and halophilic organisms, facultative and obligate halophiles, or extreme and moderate halophiles have been coined; however, these terms have not been clearly defined nor do they adequately reflect the full diversity of phenomena associated with life in the presence of salt.

The adaptations of various organisms to high salt concentrations differ qualitatively, although in all cases studied the intracellular ionic strength correlates closely to that of the saline environment. While eukaryotes, such as *Dunaliella* use photosynthetic production of glycerol to achieve internal osmotic balance with the external salinity, most prokaryotes (including halobacteria and cyanophytes) accumulate intracellular K^+ (Miller *et al.*, 1976). A specific requirement for high Na^+ that cannot be substituted by sucrose, K^+, Ca^{2+}, Mg^{2+} or other ions seems universal for all halophiles, (Mohr and Larsen, 1963; Waterbury, 1979). Enzymes, electron transport and ribosomes, once adjusted to an elevated salt concentration, lose their structural stability without it (Brock, 1969). An extreme case is that of *Halobacterium* which, unlike other prokaryotes, has a glycoprotein cell wall that is structurally dependent on Na^+ (Blaurock et al., 1976). Halobacteria can grow in saturated brines, but cease growth and lose their shape below 15% NaCl; below 12% their cell walls disintegrate (McLeod and Matula, 1962). Prokaryotes show different physiological responses to sea water: some grow well in sea water without requiring Na^+, others require Na^+ only, while specialized marine forms have a suite of additional ion requirements (McLeod, 1971; Reichelt and Baumann, 1974; Waterbury, 1979).

In addition to qualitative differences, organisms show quantitative differences with respect to the minimum requirements, growth optima and tolerance levels of the NaCl concentrations to which they can adjust. However, the criteria for distinguishing between extreme and moderate halophily and halotolerance have not been established.

Furthermore, there is variability in salt-related phenomena in the variety of saline environments inhabited, ionic compositions, concentrations and changes over time. The selective pressures in the evolution of salt-adapted organisms differ in each of these cases.

In the present paper I will address the following questions using cyanophytes as examples: (1) how to define halophily and halotolerance in terms of the survival value of these properties for cyanophytes in their natural environments; (2) are environments with constant salinity inhabited by different taxa than those with fluctuating salinity; and (3) do cyanophytes speciate along a gradient of increased salinity and evolve taxa specialized to various salinity levels?

2. Halophily or Halotolerance

Organisms living under conditions of high salinity and requiring a higher minimum concentration of NaCl than marine organisms, are usually called halophiles. The first part of this definition implies tolerance to salt, while the second part indicates a dependence on salt.

It has been recognized for half a century that a distinction should be drawn between organisms able to *withstand* high salt concentrations and those able to *grow* and *multiply* under such conditions (Hoff and Frémy, 1933). Hoff and Frémy defined as true halophiles those organisms that grow in 3M (17.55%) NaCl solution, labeling others that survive this concentration but do not grow and develop as halotolerant. The same definition was accepted by Fogg et al. (1973, p. 299). On this basis Hoff and Frémy identified several species, including *Aphanothece halophytica* Frémy as true halophiles, and called *Microcoleus chthonoplastes* Thuret (a filamentous cyanophyte known to form mats in salt works that persist throughout the full evaporation sequence) a 'typical halotolerant species'. Their isolate of *M. chthonoplastes* did not grow above 2M (11.7%) NaCl concentration. Lower limits and the minimum requirement for NaCl in *Microcoleus* have not been established.

Flannery (1956) in his definition of bacterial halophily chose 2% NaCl as the lower boundary between obligate halophiles and facultative halophiles. The latter can grow below this limit but grow better above it.

Batterton and Van Baalen (1971) studied the growth preferences of several cyanophytes isolated from marine environments and found that these particular isolates grew better at lower salinities. They could not isolate cyanophytes from brines and concluded that cyanophytes found in brines are not halophilic but merely tolerate high salinities. This view has been challenged independently by Miller et al. (1976) and by Brock (1976). Both groups isolated truly halophilic cyanophytes from brines. In addition, Brock explained that a similar 'extremely tolerant' cyanophyte isolated earlier by Kao et al. (1973) is, in fact, a true halophile, differing from his own isolate in the level of its salinity adjustment. Significantly lower salt requirements have been found in many marine cyanophytes isolated by Waterbury (1971), and yet these requirements have proved to be just as obligatory as those of the halophiles living in brines.

We can see from the preceding discussion that halophily has been defined in at least two ways. One definition is based on the various minima Na^+ concentrations required for growth. The other refers to the maximum salt concentration at which the organism can grow. Similarly, tolerance to salt has been used in two ways. One describes the range of salinities within which the organism can grow. The second is concerned with the ability of an organism to survive saline conditions although the organism does not grow.

3. Growth and Salinity

The growth of several cyanophytes isolated from saline environments and grown under various salt concentrations is summarized in Figure 1. Each of the strains presented has a characteristic range within which growth is permitted and a specific range of optimum growth. Growth can be suppressed either by salt concentrations below the minimum required or by those above the maximum that limits growth. Eukaryotic algal taxa isolated from brines have a similar spread along the salinity gradient (Gibor, 1956). Information on growth suppression by salt in marine heterotrophic bacteria indicates a similar salinity gradient distribution (McLeod, 1965). Thus, halophily cannot be defined as growth above or below any particular salinity value. It is rather a property that allows an organism to adjust its growth to a specific salinity range, which can be defined in terms of three characteristic values: minimum, optimum and maximum. Species with an optimum adjustment to high salinities also have higher minimum salt requirements and higher maxima at which growth still occurs; many of those adjusted to lower salinities may grow in freshwater, although not optimally. Those truly freshwater species with a negligible minimum Na^+ requirement also show the lowest tolerance to salt. The ranges of growth tolerance vary from taxon to taxon.

The distribution of growth optima along the salinity gradient is best described by the classical ecological scheme that relates the organisms to the limiting ecological determinant of salinity (see Ruttner, 1962). Organisms having narrow ranges of growth may be referred to as stenohaline and those having wide ranges as euryhaline (Figure 2). Extreme halophiles in the upper ranges of salinity are polyhaline; moderate halophiles, mesohaline, and those at the lower salinity ranges, such as many marine and brackish water species, may qualify as oligohaline. Combined names, e.g. oligo-euryhaline, are used to express both

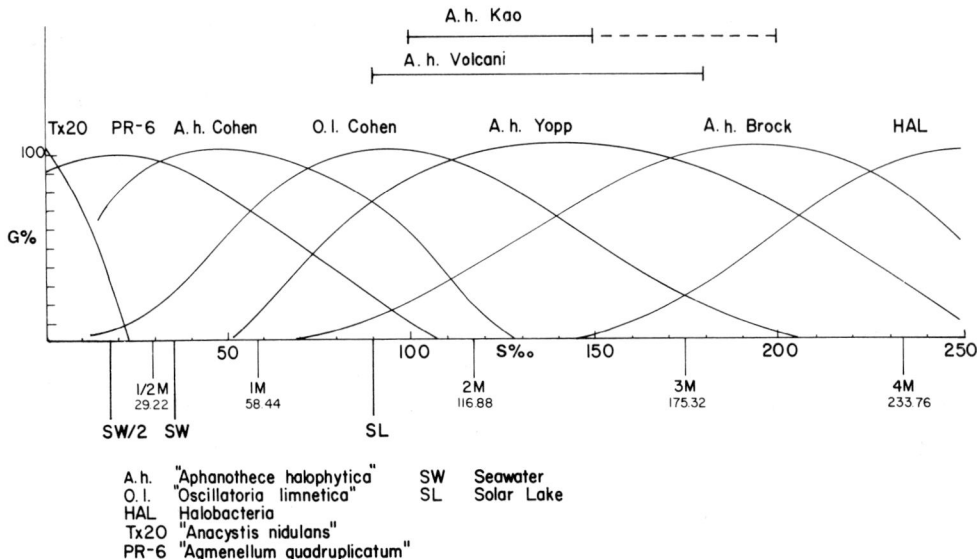

Fig. 1. Distribution of growth optima of cyanophytes isolated from environments of various salinities, expressed in percents of their optimal growth. Salinity is plotted in parts per thousand (°/oo), and in moles of NaCl (M).

range as well as optimum. This scheme restrains setting fixed boundaries to categories, stressing the gradient of increasing salinity, and thus can accommodate a large number of phenomena. It should be noted, however, that the data in Figure 1 refer to active organismal growth in 'unialgal' cultures, a condition which eliminates the effects of competition. When such organisms grow in mixed natural populations, a competitive pressure may be expected to exclude organisms from suboptimal salinity ranges, thus narrowing the width of the actual salinity niche around the optimum of each particular species.

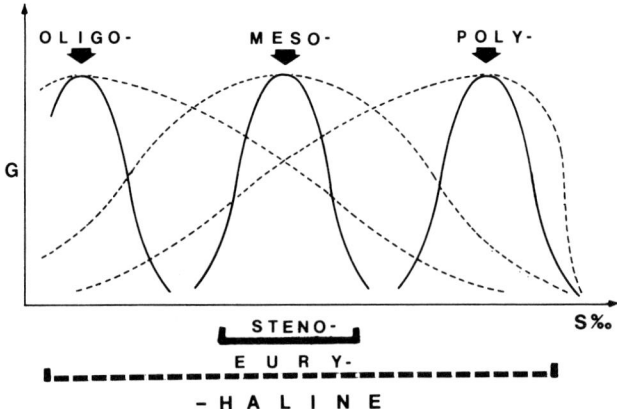

Fig. 2. Schematic presentation of the relationship of organismal adaptations to salinity. G = growth; S°/oo = salinity in parts per thousand NaCl.

4. Survival and Salinity

In the preceding section 'tolerance' to salt refers to the growth of organisms, yet the *halotolerance* of Hoff and Frémy (1933) refers to the property of withstanding, i.e. surviving, certain salinity conditions after growth has ceased. The distinction between halophilic and halotolerant organisms, arbitrarily set at 3M NaCl by these authors, is impractical in view of other criteria for halophily. It does not include moderate halophiles with lower, although obligate, salt requirements and limitations. For example, Hoff and Frémy found *Microcoleus chthonoplastes* to be halotolerant above 3M NaCl. With respect to the salt requirement and growth preference, however, the same strain might behave as a halophile at salinities below 2M NaCl concentration. Nevertheless, Hoff and Frémy's definition of halotolerance is important because it stresses *survival* under high salinity conditions.

Specific differences in the ability to survive high salinity have been shown by Stewart (1964) who compared at various salinities, the growth of two cyanophytes isolated from the marine environment. The taxa he compared, *Calothrix scopulorum* and *Nostoc endophytum*, cease growth at their respective salinity maxima, but only *Calothrix* resumes growth when lower salinities are re-established. Thus *Calothrix* shows a reversible growth inhibition.

While the growth of an organism may be confined to wider or narrower ranges of salinity (eury- vs steno), its ability to survive adverse salinity conditions may stretch far beyond the conditions that permit active growth. Thus, growth and survival represent two independent properties. The selective advantage of the ability to survive adverse conditions is obvious in an environment with fluctuating salinity. For example, an organism which can survive a wide range of salinities, but grows actively only within a narrow range, may dominate such habitats as the intertidal and wave spray zones of the sea coast by alternating short spurts of growth with long periods of dormancy, while its competitors are being eliminated by one or the other extreme of the fluctuating range. Therefore, the occurrence of an organism and its dominance in a natural habitat may reflect either its ability to survive or its ability to grow under the prevailing conditions of that particular habitat.

Survival of organisms or of their dormant stages under conditions not permitting their growth is a phenomenon which accounts for the common enrichment, isolation and culturing of organisms alien to the habitat under study, while the culturing of indigenous organisms fails. This may explain, for example, why Batterton and Van Baalen obtained no halophiles from brines, why some halophiles were isolated from terrestrial environments, and why many bacteria isolated from marine environment behaved as non-marine organisms (McLeod, 1965). However, the abundance or dominance of an organism in the habitat implies also an ability to grow and compete for this habitat within certain recurring periods when growth is permitted. Therefore, the presence or dominance of an organism in its natural habitat should be noted at the time of its isolation in culture.

5. Saline Environments and Their Inhabitants

Both halophily and halotolerance constitute different evolutionary strategies and thus offer different selective advantages. Accordingly, we can predict that stable environments with constant salinity at any concentration will select for halophily and stenohaline specialization. Conversely, environments with fluctuating salinity will select for halotolerance and euryhalinity. We have observed that stable and fluctuating environments are inhabited by different taxa as will be exemplified in the following discussion.

The open ocean is probably the most constant saline environment, while the rockpools and wave spray zone of its coasts are probably the least constant ones. Lagoons, tidal flats, salt marshes, and estuaries are intermediate. In arid regions where evaporation rates exceed atmospheric precipitation, water bodies of various sizes secluded from the ocean, embayments, lagoons, salt ponds gradually evaporate. In the process, the concentration of solutes increases, creating an array of environments with elevated salinities. Under humid climatic conditions, such secluded embayments are gradually diluted by freshwater resulting in a similar range of low salinity environments. The relative constancy of salinity depends on the size of a bounded saline environment. Salinity fluctuation in a semi-bounded water body is regulated by exchange with the ocean waters.

The tropical open ocean is populated by a few unique cyanophyte taxa, which form blooms of considerable population density. The genus *Trichodesmium* is the best known (Figure 5A). The nitrogen fixation in natural populations of *Trichodesmium* has been studied although pure cultures have not been achieved, indicating that the growth requirements of *Trichodesmium* are complex and the niche specialization, narrow. Fluctuating environments in the intertidal zone harbor epilithic and endolithic (carbonate boring) species on rocky shores, many of which can be distinguished from freshwater forms on morphological grounds. These have a wide distribution, although some regionality and latitudinal zonation in cyanophytes and microalgae has been observed (LeCampion, 1970; Schneider, 1976; Golubic et al., 1975). A different set of microorganisms, primarily cyanophytes, is found in sand and mudflats; several of these taxa form laminated stromatolitic structures analogous to fossil stromatolites, particularly to those of Precambrian age (Golubic, 1976a, 1976b; Golubic and Hofmann, 1976). Microbial communities similar to those occurring in the intertidal flats of tropical lagoonal environments have been noted in salt works which are active during a short season in the summer. All salt pans used in the salt works of Piran, Yugoslavia, maintain a salinity close to that of the sea water throughout the winter. Only during 6 to 8 weeks in the summer do a series of these pans contain water of increasing salinity up to the precipitation of halite (Herrmann et al., 1973).

Relatively constant salinity is maintained in large ponds that are adjacent to the ocean waters as well as in large secluded water bodies and inland lakes of arid regions, and all known halophilic cyanophytes have been isolated from such environments: the Great Salt Lake (Brock, 1976), a solar evaporation pond, Redwood City California (Yopp and Tindall, 1974), a salt pond in Puerto Rico (Kao et al., 1973), the Dead Sea (Elazari-Volcani, 1944), Solar Lake, Sinai (Cohen, 1975), or isolated from salt that precipitated in these environments (Hoff and Frémy, 1933).

We have conducted field studies and investigated the occurrence and dominance of cyanophytes in a series of hypersaline ponds at Yallahs, south Jamaica and of the Solar Lake, Sinai, near Eilat, Israel.

Fig. 3. View of the Yallahs salt ponds, south Jamaica. Great Salt Pond is to the left and Little Salt Pond to the right, Caribbean Sea in the background.

The Yallahs ponds, Jamaica (Figure 3) are three shallow hypersaline basins, Great Salt Pond, Little Salt Pond and West Pond separated from the ocean by a barrier beach and mutually connected by narrow channels. At the beginning of the dry season (February, 1978) the salinity in these ponds was 2.2M, 2.5M, and 4.5M NaCl. Both Great and Little Salt Ponds had localized seepage areas of lower salinity (0.54M) where sea water entered th ponds in the form of 'springs'. In addition to these seepage areas, small temporary puddles along the shores receive periodic windblown pond water, which either evaporates or becomes diluted by rain. These seepage locations and littoral puddles have the highest salinity fluctuation. In the course of the dry season the salinity gradually increases in all three ponds, but only the West Pond evaporates completely.

The Solar Lake, (Sinai peninsula) is smaller and deeper than the Yallahs Ponds. Seepage from the adjacent Red Sea occurs profusely through a beach barrier. The fluctuation of the water level in this pond is a function of the ratio of water seepage rate to the evaporation rate, as rain rarely occurs in this desert region. Seepage of lighter sea water during the winter results in overlayering of the heavier brine water of the Solar Lake forming one or more chemoclines. During the summer, evaporation causes a salinity increase at the surface ultimately resulting in mixing of the entire water column at the end of the summer (Cohen *et al.*, 1977).

Within these hypersaline ponds, three distinct habitats could be identified which are occupied by different microbial communities: (1) the seepage littoral community is

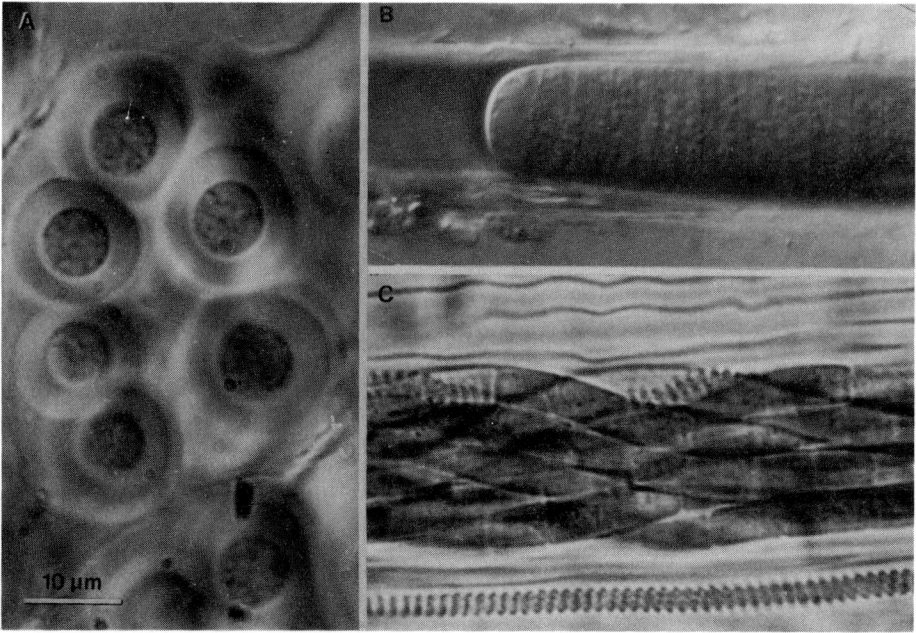

Fig. 4. Cyanophytes that dominate environments with fluctuating salinities (intertidal zone of tropical lagoons). (A) *Entophysalis major* Erceg. (B) *Lyngbya aestuarii* (Mertens) Liebman; (C) *Microcoleus chthonoplastes* Thuret with *Spirulina subtilissima* Kuetz. (Migrated into the *Microcoleus* sheath.) Scale bar = 10 µm.

composed of organisms that are in contact with varying salinity and form mats; (2) the pond plankton refers to organisms which are suspended in the water or loosely settled on the bottom; and (3) the pond benthos community refers to organisms firmly attached to the bottom or coating gypsum crusts and stones.

(1) The microbial community of the seepage areas forms typical algal mats of firm, gelatinous to leathery consistency. It is dominated at the primary producer level by the cyanophytes *Entophysalis major*, *Lyngbya aestuarii* and *Microcoleus chthonoplastes* (Figures 4A, B, and C). Other cyanophytes such as *Spirulina tenerrima*, *S. subsalsa* (Figures 4C, 5N), *Johannesbaptistia pellucida* and various species of *Schizothrix* form a minor component of these mats. The microorganisms inhabitating seepage areas are not significantly different from those that dominate intertidal flats and temporary saline ponds of tropical and subtropical coasts or those found in salt works with fluctuating salinity (e.g. Piran, Yugoslavia (Golubic, 1973, 1976)).

(2) The main water body of the shallow Yallahs ponds is mixed daily and contains dense populations of suspended microorganisms. During calm periods some of these organisms settle loosely on the bottom and are resuspended with the onset of wind. These suspended organisms give the ponds their characteristic color (Figure 3). The Great Salt Pond has a brownish-green color caused by a dominant coccoid cyanophyte (Figure 5B, C). This yet undescribed organism has unique properties which place it intermediate

to the genera *Aphanocapsa* and *Coelospherium*. Small spheroid cells (1.5–2 μm in diameter) divide in two planes but remain attached by the thin envelopes they shed, forming, hollow spherical colonies. Little Salt Pond and West Salt Pond were orange-pink during the time of study. The dominant planktonic organism was an equally unusual, new coccoid cyanophyte morphologically close to the genera *Dactylococcopsis* and *Synechococcus* (Figures 5G, H, J, K). The shape of its extremely long cells (up to 300 μm long) is a thin spindle (3.2 μm wide), which, when fully differentiated, has attenuated ends. Cell division takes place transversally and starts with a constriction midway along the cell (Figures 5G, H). The cytoplasm of these cells is arranged in distinct spirals around the periphery of the spindle; small highly refractive granules which could be gas vesicles are often seen within the spiral bands of cytoplasm (Figure 5J). Another small species of the same genus, not yet described, occurs in the Solar Lake.

Minor components of the plankton community in all three Yallahs ponds are several taxa of coccoid cyanophytes with highly 'keratomized' cytoplasm (Figures 5D–F). Keratomization involves a reticulate separation of cytoplasm and a vacuole-like space. This phenomenon is not pathological (Geitler, 1930). These new coccoid cyanophytes are similar in morphology to the genera *Synechocystis* and *Synechococcus*. Their spheroid, isodiametric shape is consistent with the description of *Synechocystis*. However, the cells divide in one plane (see Figure 5F), which is a characteristic of *Synechococcus*. A eukaryote commonly encountered in Yallahs ponds in a colorless flagellate with four anterior flagella (*Tetramitus salinus*? Figures 5I, J).

(3) Benthic micorbial populations either form crusts that adhere firmly to hard substrates such as stones or gypsum crystals or coat the bottom with soft, flocculent mats. Crusty colonies adhering firmly to hard substrate in Yallahs ponds are formed by a pleurocapsacean cyanophyte similar to the genus *Xenococcus* (Figure 5R). Another common constituent of the Yallahs pond benthos is a soft mat forming *Phormidium*, characterized by keratomized cytoplasm, and quite similar to the *Phormidium* sp. isolated by Hoff and Frémy (1933, Figure 2) from salt crusts. At the time of this study the benthic cyanophyte cover in Yallahs ponds was very sparse, although *Phormidium* did form coherent mats during previous collection visits (A. Weiss, personal communication, 1978) (Figures 5P, Q).

An extremely dense population of brine flies, their larvae and puppae was present in the ponds in February, suggesting that brine fly larvae may have overgrazed algal mats as Brock *et al.* (1969) postulated for thermal springs.

The most conspicuous benthic cyanophyte population of the 4–5 m, deep Solar Lake, Sinai, is the flocculent, dark blue-green mat composed of several strains of a halophilic oscillatoriacean of genus *Phormidium* (Campbell and Golubic, unpublished results). One of these strains isolated in culture under the provisional name *Oscillatoria limnetica* was shown to be halophilic, and, in addition, to have the unusual property of facultative anaerobic photosynthesis using H_2S as a hydrogen donor (Cohen *et al.*, 1975).

Aphanothece is the most common genus that forms mucilagenous coatings in the benthos of many salt ponds. Examples of two forms of *Aphanothece* that coexist in Yallahs ponds are shown in Figures 5L, M. Two larger *Aphanothece* coexisting in Solar

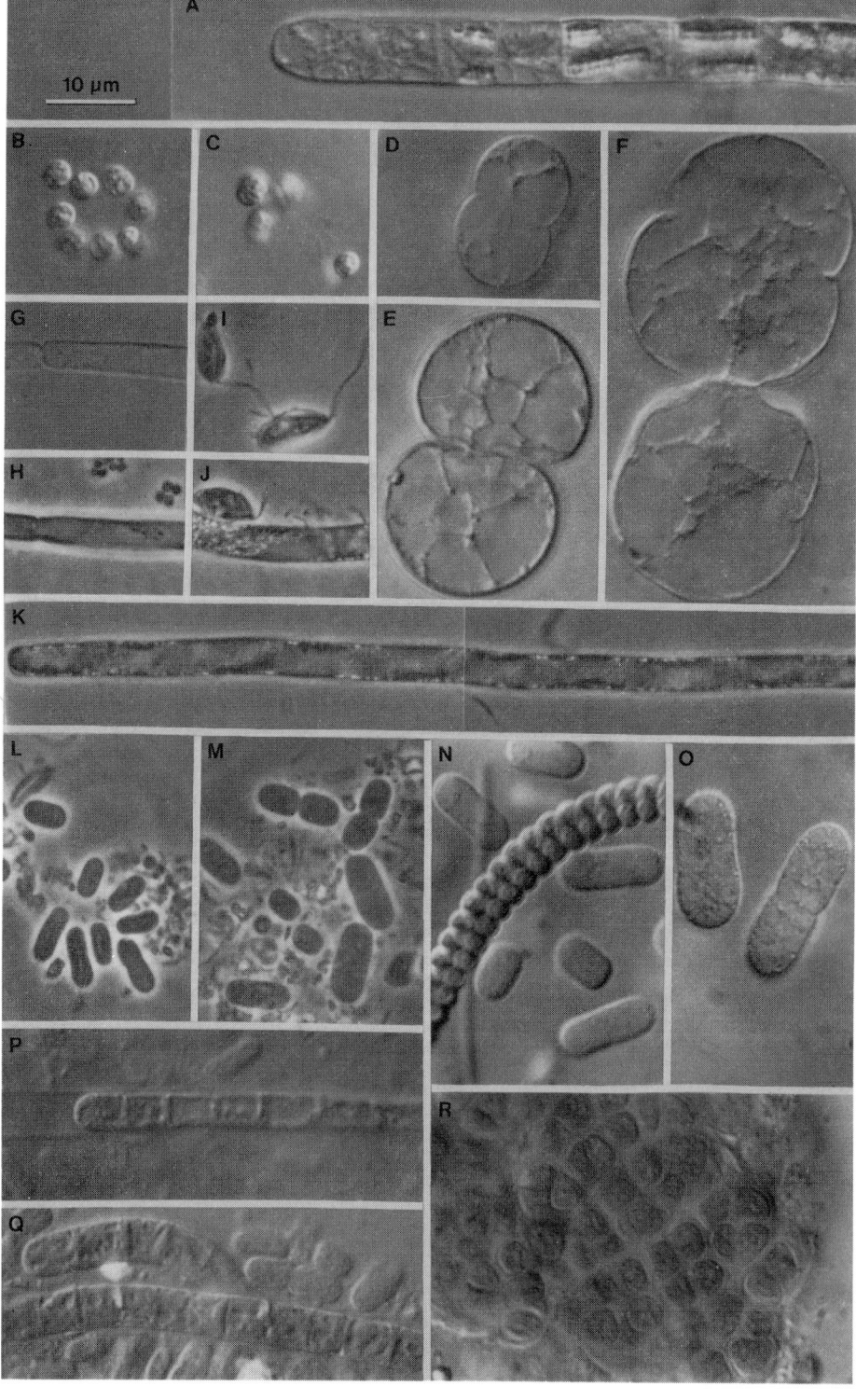

Pond are shown in Figures, 5N, O. *Aphanothece* of various sizes occur in Yallahs ponds, on slopes and margins of the Solar Lake, in the Great Salt Lake, in ponds in the upper intertidal zone of the Arabian Gulf coast and in subtidal of the lagoons of tropical seas. Several strains isolated in culture have specifically different halophilic properties as shown in Figure 1. The existence of morphologically, physiologically and ecologically distinct populations of *Aphanothece* in saline environments raises the question of their species identity: Is it scientifically justified to call them all *Aphanothece halophytica*?

6. Cyanophyte Speciation in Saline Environments

The discovery of unusual organisms with unique properties that enable their growth and survival in saline environments raises the following, taxomically relevant questions: Do cyanophytes speciate along a salinity gradient and what is the best taxonomic reference for naming and comparing cyanophytes from saline environments?

So far we have established that the total number of discernible taxa decreases with the increase of salinity, that as the salinity increases, the species composition shifts, and that environments with constant and with fluctuating salinity are inhabited by different organisms. We have further established that the properties which characterize halophilic and halotolerant organisms are adjusted within a limited, genetically determined, range of tolerance to salt. In addition to a particular adjustment to NaCl, these organisms have other mutually correlated properties (growth requirements, cytological properties, morphology) that distinguish them from similar organisms in other environments.

From these considerations the first question can be answered affirmatively: cyanophytes did evolve specialized taxa which acquired adaptations that permit their survival, growth, and dominance in various saline environments. Thus halophilic, halotolerant and marine cyanophytes represent separate taxa that can be distinguished from each other and from similar forms in freshwater and terrestrial environments.

The second question concerns the nomenclatorial reference for these taxa and requires that the hierarchical level of taxonomic entities be established. This is more difficult as

---

← Fig. 5. Cyanophytes that dominate environments with constant salinity: Tropical open ocean (A), Great Salt Pond (B–D) and Little Salt Pond (E–M, P–R), Jamaica and Solar Lake (N, O) Sinai. Scale bar for all organisms is 10 μm long (upper left). (A) *Trichodesmium thiebautii* Gom. from the Atlantic Ocean north of Puerto Rico; note the cytoplasmic reticulation and packets of gas vesicles; (B–C) *Coelosphaerium* sp., cells are held together by discarded envelopes forming hollow sphere colonies; (D–F) *Synechococcus* sp., characterized by strong reticulation (keratomization) of the cytoplasm, and a cell division in one plane (1D), the globular cell shape is similar to *Synechocystis* (2D); (G, H, J, K) *Dactylococcopsis* sp., with long spindle-shaped cells that divide transversally by constriction (G, H), cytoplasm spirally arranged with refractive inclusions (J, K); (I, J) *Tetramitus* sp. ?; (L, M) *Aphanothece* sp. from Yallahs pond benthos (phase contrast); (N, O) *Aphanothece* sp. from Solar Lake (Nomarski contrast) with *Spirulina subsalsa* Oerst. intermingled (N), and a larger taxon from the littoral of the Solar Lake (O); (P, Q) *Phormidium* sp., coating benthos of Yallahs ponds, intermixed with *Aphanothece* sp.; (R) *Xenococcus* sp. forms hard crust on gypsum crystals in Little Salt Pond of Yallahs.

it is dependent on the very concept of a species in prokaryotes, and to date consistent criteria have not been used.

Bacteriological classification is based on cultured strains, i.e. clonal lineages with uniform genetic properties. A strain is comparable to a tissue culture derived from an individual and is representative of a *very* small sample of a natural population. In bacteriology a species represents a group of strains that share a certain unspecified number of traits.

Taxonomic practice in ecology should not be based on individuals or clones, but rather on frequency distributions and combinations of traits that determine the establishment and continuity of a population in its natural environment. The boundaries of a species should be derived from a composite view of surviving populations of common descent that perform the same range of ecological functions, i.e. that occupy the same ecological niche.

Several criteria can be employed in determination and characterization of species. First, discernible properties should be recorded, evaluated and, if possible, quantified at the population level, using morphometry and statistics. Second, these properties should not be assessed separately but correlated as combinations, thus increasing the confidence in identifying individuals and populations with common properties. Third, a population so described should be compared with published species descriptions and oriented within the system of the basis of similarities and differences to several related taxa. Fourth, Natural populations are rarely identical with those described as types, and the degree of difference encountered can be defined as a 'distance' from a point of reference. If more than 3 properties are simultaneously evaluated, a conceptual framework of taxonomic hyperspace may be employed (Sneath and Sokal, 1973), in which populations separate as clusters. Fifth, a need for a description of a new species arises only if the newly identified population shows major differences from all described species, which constitutes a gap in the hyperspace reference matrix.

The problem of determination of species and genera is cyanophytes is exemplified by a group of rod-shaped unicells which are common in saline environments. A number of isolates of such organisms (Figure 1) have all been assigned to one species: *Aphanothece halophytica* Frémy. Occurrence of a single species across a wide range of salinites would imply either an extreme euryhalinity of a single, functionally adaptible taxon with capacity for reversible accommodation to salt, or that the species has been defined too broadly and contains physiologically and genetically distinct and stable taxonomic subunits. These should be characterized and named separately.

Attempts to change halophilic properties of at least one of these strains have failed (Tindall *et al.*, 1978), and some of the strains show differences other than growth optima with respect to salinity. Growth optima of cyanophytes from saline environments summarized in Figure 1 show a distribution along the salinity gradient analogous to that of tree growth on the slopes of Smoky Mountains (Whittaker, 1965), where dominance of tree species reflects their optimal adaptation to average temperature and availability of water. Thus, the halophilic cyanophytes behave in a way analogous to distinct species of plants. The *Aphanothece* illustrated in Figures, 5L–O show significant differences in shapes and sizes with little variation within each subpopulation. However, it was also reported that size and shape change with changing culture conditions (Brock, 1976; Yopp et al., 1978). Whether the taxa under discussion should be grouped in one or more species remains inconclusive. A correlation of both morphological and physiological properties in a cluster analysis is needed in order to resolve the species identity of these forms.

The question of genus assignment of these forms has also been a subject of controversy. In their study of halophilic cyanophytes, Hoff and Frémy (1933) were confounded by the morphological variability of unicellular cyanophytes grown in high salt concentration media and, particularly, by the presence of the highly irregular 'giant cells' in their cultures. They could not agree on the taxonomic identity of the forms observed: Frémy distinguished between *Aphanocapsa litoralis* Hansgirg and a newly described species *Aphanothece halophytica* Frémy, while Hoff considered all coccoid forms observed to be variants of the former species. Brock (1976) used the name *Aphanothece halophytica* 'for simplicity' to describe his isolate from the Great Salt Lake and then used the name *Aphanocapsa halophytica* interchangeably (probably as a compromise between *Aphanocapsa litoralis* and *Aphanothece halophytica*).

Genera in coccoid cyanophytes are defined on the basis of cell division patterns, cell shape, production of extracellular envelopes and colony formation (Geitler, 1930; Golubic, 1976b). The genus *Aphanothece* (as well as *Synechococcus* and *Gloeothece*) is characterized by division of usually rod-shaped cells in a single plane, while the genus *Aphanocapsa* (as well as *Synechocystis* and *Gloeocapsa*) by division of usually spherical cells in two or three planes. These cell division patterns are constant within each taxon and persist until cell division is halted by excess salt (Schiewer and Jonas, 1977a, b), although they can be masked by pathological cell deformation under these conditions.

Thus, all rod-shaped unicellular cyanophytes that divide in one plane and are embedded in an amorphous mucilage belong to the genus *Aphanothece* (Figures 5L–O), while those suspended singly and without coherent mucilage are classified as the genus *Synechococcus*. Both genera are common in saline environments. Some round-celled forms which also divide in one plane (Figures 5D–F) should also be classified within the latter genus. Long, transversely dividing rods with pointed or curved ends should be classified within the genus *Dactylococcopsis*.

In conclusion, all forms under discussion, which have been classified as *Aphanothece halophytica* belong to the same genus, but probably represent separate species. The halophily has been demonstrated to date only in a few cyanophytes which were isolated in cultures. These represent a small fraction of the unique cyanophyte assemblages that coexist and dominate in environments of high salinity. It is predicted that these taxa once isolated in culture, will prove to have similar halophilic properties.

Acknowledgements

I thank A. Wais, Tufts University and University of West Indies for collaboration and support of field work, and for providing research and living facilities in Jamaica, as well as for information on the physical and chemical properties of Yallahs ponds; Y. Cohen,

Heinz Steinitz Marine Biological Laboratory, Eilat, Israel for his help in field work in Solar Lake. F. Motzkin and M. Potts provided the use of their facilities and helped in transportation. J. Waterbury, Woods Hole Oceanographic Institution, helped with the literature survey and provided valuable criticism. S. E. Campbell, Boston University, assisted in field work at both sites, critically read the manuscript and gave valuable suggestions.

The research was supported by the NSF Grants DEB 76-21542-AO1 to A. Wais, and OCE 12999-AO2 to S. Golubic and EAR 76-84233-AO1 to S. Golubic and B. Cameron.

References

Batterton, J. C. and Van Baalen, C.: 1971, *Arch. Microbiol.* 76, 151.
Blaurock, A. E., Stoeckenius, W., Oesterhelt, D. and Scherphof, G. L.: 1976, *J. Cell Biol.* 71, 16.
Brock, T. D.: 1969, *Symp. Soc. Gen. Microbiol.* 19, 15.
Brock, T. D.: 1976, *Arch. Microbiol.* 107, 109.
Brock, M. L., Wiegert, R. G., and Brock, T. D.: 1969, *Ecology* 50, 192.
Campbell, S. E.: 1980, Ph. D. Thesis, Boston Univ., Boston.
Cohen, Y.: 1975, Ph.D. thesis, Hebrew Univ., Jerusalem.
Cohen, Y., Padan, E., and Shilo, M.: 1975, *J. Bacteriol.* 123, 855.
Cohen, Y., Krumbein, W. E., Goldberg, M., and Shilo, M.: 1977, *Limnology and Oceanography* 22, 597.
Elazari-Volcani, B.: 1944, *Papers to Commemorate the 70th Anniversary of Dr. Ch. Weizmann*, Daniel Sieff Research Inst. Rehovoth, Israel, pp. 71–85.
Flannery, W. L.: 1956, *Bacter. Review* 20, 49.
Fogg, G. E., Stewart, W. D. P., Fay, P., and Walsby, A. E.: 1973, *The Blue-Green Algae,* Academic Press, London–New York, 459 pp.
Geitler, L.: 1930, in Rabenhorst (ed.), *Kryptogemenflora*, Akad. Verlagsgesellsch. Leipzig, Johnson Reprint Corp. New York–London, 1971, 1196 pp.
Gibor, A.: 1956, *Biol. Bull.* 111, 223.
Golubic, S.: 1973, in N. Carr and B. A. Whitton (eds.), *The Biology of Blue-Green Algae,* Blackwell Scientific Publ., Oxford, pp. 434–472.
Golubic, S.: 1976a, in M. R. Walter (ed.), *Stromatolites. Developments in Sedimentology* 20, Elsevier Scientific Publishing Co., Amsterdam–Oxford–New York, pp. 113–126.
Golubic, S.: 1976b, in M. R. Walter (ed.), *Stromatolites. Developments in Sedimentology* 20, Elsevier Scientific Publishing Co., Amsterdam–Oxford–New York, pp. 127–140.
Golubic, S. and Hofmann, H.: 1976, *J. Paleontol.* 50, 1074.
Golubic, S., Perkins, R. D., and Lukas, K. J.: 1975, in R. W. Frey (ed.), *The Study of Trace Fossils* Springer Verlag, New York–Heidelberg–Berlin, pp. 229–259.
Herrmann, A. G., Knake, D., Schneider, J., and Peters, H.: 1973, *Contr. Mineral. Petrol.* 40, 1.
Hoff, T. and Frémy, P.: 1933, *Red. Trav. Bot. Neerl.* 30, 140.
Kao, O. H. W., Berns, D. S., and Town, W. R.: 1973, *Biochem. J.* 131, 39.
Le Campion-Alsumard, T.: 1970, *Schweiz. Z. Hydrol.* 32, 552.
McLeod, R. A.: 1965, *Bacteriol. Rev.* 29, 9.
McLeod, R. A. and Matula, T. I.: 1962, *Can. J. Microbiol.* 8, 883.
Miller, D. M., Jones, J. H., Yopp, J. H., Tindall, D. R., and Schmid, W. E.: 1976, *Arch. Microbiol.* 111, 145.
Mohr, V. and Larsen, H.: 1963, *J. Gen. Microbiol.* 31, 267.
Reichelt, J. L. and Baumann, P.: 1974, *Arch. Microbiol.* 97, 329.
Ruttner, F.: 1962, *Grundriss der Limnologie,* 3rd ed., Walter de Gruyter and Co., 332 pp., (English transl., Univ. of Toronto Press, 1963, 295 pp.).
Schneider, J.: 1976, *Contr. Sedimentology* 6, 1.

Sneath, P. H. A. and Sokal, R. R.: 1973, *Numerical Taxonomy*, Freeman and Co., San Francisco, 573 pp.
Schiewer, U. and Jonas, L.: 1977a, *Arch. Protistenk.* 119, 127; (German w. English summary).
Schiewer, U. and Jonas, L.: 1977b, *Arch. Protistenk.* 119, 146.
Stewart, W. D. P.: 1964, *J. Gen. Microbiol.* 36, 415.
Tindall, D. R., Yopp, J. H., Miller, D. M., and Schmid, W. E.: 1978, *Phycologia* 17, 179.
Waterbury, J. B.: 1971, Ph. D. thesis, Univ. of California, Berkeley, 279 pp.
Waterbury, J. B.: 1979, in *The Prokaryotes*, Springer Verlag, in press.
Whittaker, R. H.: 1965, *Science* 147, 250.
Yopp, J. H. and Tindall, D. R.: 1974, *ASB Bull.* 21, 87.
Yopp. J. H., Tindall, D. R., Miller, D. M., and Schmid, W. E.: 1978, *Phycologia* 17, 172.

SOIL STABILIZATION BY A PROKARYOTIC DESERT CRUST: IMPLICATIONS FOR PRECAMBRIAN LAND BIOTA

S. E. CAMPBELL

Boston University Department of Biology Boston, Mass. 02215, U.S.A.

Abstract. A cyanophyte dominated mat, desert crust, forms the ground cover in areas measuring hundreds of square meters in Utah and smaller patches in Colorado. The algal mat shows stromatolitic features such as sediment trapping and accretion, a convoluted surface, and polygonal cracking. Sand and clay particles are immobilized by a dense network of filaments of the two dominating cyanophyte species, *Microcoleus vaginatus* and *M. chthonoplastes*, which secrete sheaths to which particles adhere. These microorganisms can tolerate long periods of desiccation and are capable of instant reactivation and migration following wetting. Migration occurs in two events: 1. immediately following wetting of dry mat, trichomes are mechanically expelled from the sheath as it swells during rehydration, and 2. subsequently, trichomes begin a self-propelled gliding motility which is accompanied by further production of sheath. The maximum distance traveled on solid agar by trichomes of *Microcoleus vaginatus* during a 12 hour period of light was 4.8 cm. This corresponds to approximately 500 times the length of the fastest trichome, and provides a measure of the potential for spreading of the mat in nature via the motility of the trichomes.

Dehydration resistance of the sheath modifies the extracellular environment of the trichomes and enables their transition to dormancy. Following prolonged wetting and evaporative drying of the mat in the laboratory, a smooth wafer-like crust is formed by the sheaths of *Microcleus* trichomes that have migrated to the surface. Calcium carbonate precipitates among the algal filaments under experimental conditions, indicating a potential for mat lithification and fossilization in the form of a caliche crust. It is suggested that limestones containing tubular microfossils may, in part, be of such an origin.

The formation of mature Precambrian soils may be attributable to soil accretion, stabilization, and biogenic modification by blue-green algal land mats similar to desert crust.

1. Introduction

Stabilization of some soils in the arid regions of the United States of America results from sediment trapping and accretion by blue-green algal communities. They form stromatolitic mats on land that are known as 'desert crust'. Desert crust (see Figure 1) is geographically widespread in the southwestern United States, having been reported in Oklahoma, Kansas, Texas (Booth, 1941), New Mexico (Shields *et al.*, 1957), and Idaho (Brock, 1975). This study reports new occurrences in Utah and Colorado.

A network of filamentous cyanophytes is responsible for consolidation of sand, and is frequently one to three inches deep. The crust may rest on rock (see Figure 2) or soil (see Figure 3) in areas of flat topography. It is absent from erosional channels and steeply sloping terrain. The mat's surface is typically rough and undulating, and is rigid, although the mat is fragile enough to disintegrate into sand when pressure is applied. Some polygonal cracking occurs which allows careful removal of intact chunks of the crust for structural analysis such as scanning electron microscopy (see Figures 4 and 5). The ground cover afforded by desert crust varies from meter square isolated patches on a Colorado grassland plateau to areas of several hundred square meters which occur throughout Canyonlands National Park in Utah.

Soil algae have been shown to be instrumental in soil binding and water retention (Booth, 1941) and to be important sources of fixed nitrogen (Shields *et al.*, 1957). In this study the stromatolitic nature of a modern terrestrial algal mat is recognized and observations are made on the species composition of the algal community, the adaptations of the algae to water shortage which prevails during extended periods of time in the desert, and their response to wetting and drying.

Blue-green algal stromatolites occurred in the marine subtidal and intertidal zones during the Precambrian, and continue to occupy similar niches in the Recent (e.g. Shark Bay and the Persian Gulf), and some of the Precambrian stromatolite-dwelling organisms have been shown to have modern morphological counterparts (Golubic, 1976; Golubic and Hofmann, 1976).

This paper discusses the evolutionary origins of desert crust algal mats and their possible role in the formation of Precambrian soils.

2. Materials and Methods

Pieces of dormant desert crust were collected dry in the field and stored near a window in an effort to simulate natural conditions of light. Mat placed in finger bowls was dampened with distilled water in order to achieve a laboratory simulation of the mat's response to rain. It was then allowed to dry gradually and completely.

Dry cyanophyte filaments were selectively plucked out of the natural field-collected mat for microscopic preparations using forceps. Small chunks of crust were wetted and transferred carefully through a series of droplets of distilled water to clean organisms of adhering sand particles, to enable photomicrography and identification of species.

Light microscopy and photomicrography were accomplished using a Zeiss Universal microscope with transmitted light, Nomarsky interference contrast, and phase contrast illumination (see Figures 6–15). Scanning electron microscopy was done on an AMR model 1000 SEM.

Fig. 1. Desert crust covers large areas in Canyonlands National Park, Utah. Fig. 2. Desert crust forming over flat rock. Fig. 3. Desert crust forming on sand. The mat can be several inches deep, and has a brittle, convoluted surface. Fig. 4. SEM of the experimentally wetted and dried crust shows a horizontal surface layer of cyanophyte filaments. Dried moss gametophytes are above, and sand grains are below. All scales shown indicate micrometers. Fig. 5. Beginning calcite precipitation in the surface layer of Figure 4. Fig. 6. Cross section of a large filament of *Microcoleus* which was resin-embedded, thin sectioned, and stained. More than forty trichomes are in a common sheath, which is actually a composite of smaller bundles that appear here as darkened rings. During dehydration, shrinkage of sheath material causes tubular spaces (arrow) to form around the trichomes. Fig. 7. Bundled trichomes of *Microcoleus vaginatus* (Vaucher) Gomont in a common sheath. Note the capitate, narrowing trichome tip characteristic of this species. All light photomicrographs are with Nomarski contrast illumination. Fig. 8. *Microcoleus chthonoplastes* Thuret. Trichomes end with a bullet-shaped end cell.

Study of algal motility was conducted as follows: cyanophytes were tweezed from the dry mat and put on a dry slide. A coverslip was placed on top, and the microscope focused to 160x in order to best observe motility of entire trichomes. A drop of distilled water was touched to the edge of a coverslip, and by capillary action drawn beneath it. Maximum rates of gliding motility were measured directly by observation with the microscope and timing with a stopwatch. Long term migration, was followed by recording the position of trichomes on an agar surface at time intervals. This approach is a more valid basis for extrapolation to rates of migration in nature than using the maximum rates of gliding described above, because direct observation of trichomes showed that they move intermittently and at varying speeds.

The migration of trichomes over a period of several hours was observed by placing single dry filaments (each containing ca. 50–100 trichomes) at the center of twenty seven solid agar plates (BG-11 freshwater medium, see Stainer *et al.*, 1971, with 1.5% agar) under the following conditions: (1) the inoculation was made on seven plates in the dark and after 12 hours of darkness each plate was illuminated for 12 hours; (2) the inoculation was made on twenty plates in the light and after 12 hours of light the plates were placed in darkness for 12 hours. During periods of illumination plates were observed several times using a dissecting microscope. Plates kept in darkness were observed only at the end of the dark period. The migrating trichomes leave visible trails along the agar surface (as long as it is fairly solid). The lengths of the trails were measured (from the leading trichome to the point of the inoculum) and the average motility rate of that trichome was calculated in $\mu m\ sec^{-1}$.

Bundled trichomes were embedded and sectioned for light microscopy. However, the embedding method described below also provides a specimen preparation suitable for transmission electron microscopy. When trichomes migrate on agar they tend to move together in streams. Streams were fixed with glutaraldehyde (2.5%), covered with a drop of warm liquid agar, and a block of agar containing the trichomes was cut out of the plate (see Rippka *et al.*, 1974). This block was then stained with OsO_4, transferred through cacodylate buffer and then a dehydration series of gradually increasing ethanol concentration, placed in propylene oxide, and then a 1:1 mixture of propylene oxide and liquid epon 812 resin. The block was then transferred twice into pure epon before curing in an oven at 60°C for three days. The resulting hard plastic block was then sectioned on an M-1 microtome, cutting cross-sections through a *Microcoleus* filament. The thin sections were stained with toluidine blue (see Figure 6).

3. Results and Discussion

3.1. THE SPECIES COMPOSITION OF THE ALGAL MAT COMMUNITY

The Utah and Colorado desert crust microbial communities are predominantly prokaryotic. Two species of filamentous cyanophytes, *Microcoleus vaginatus* (Vaucher)

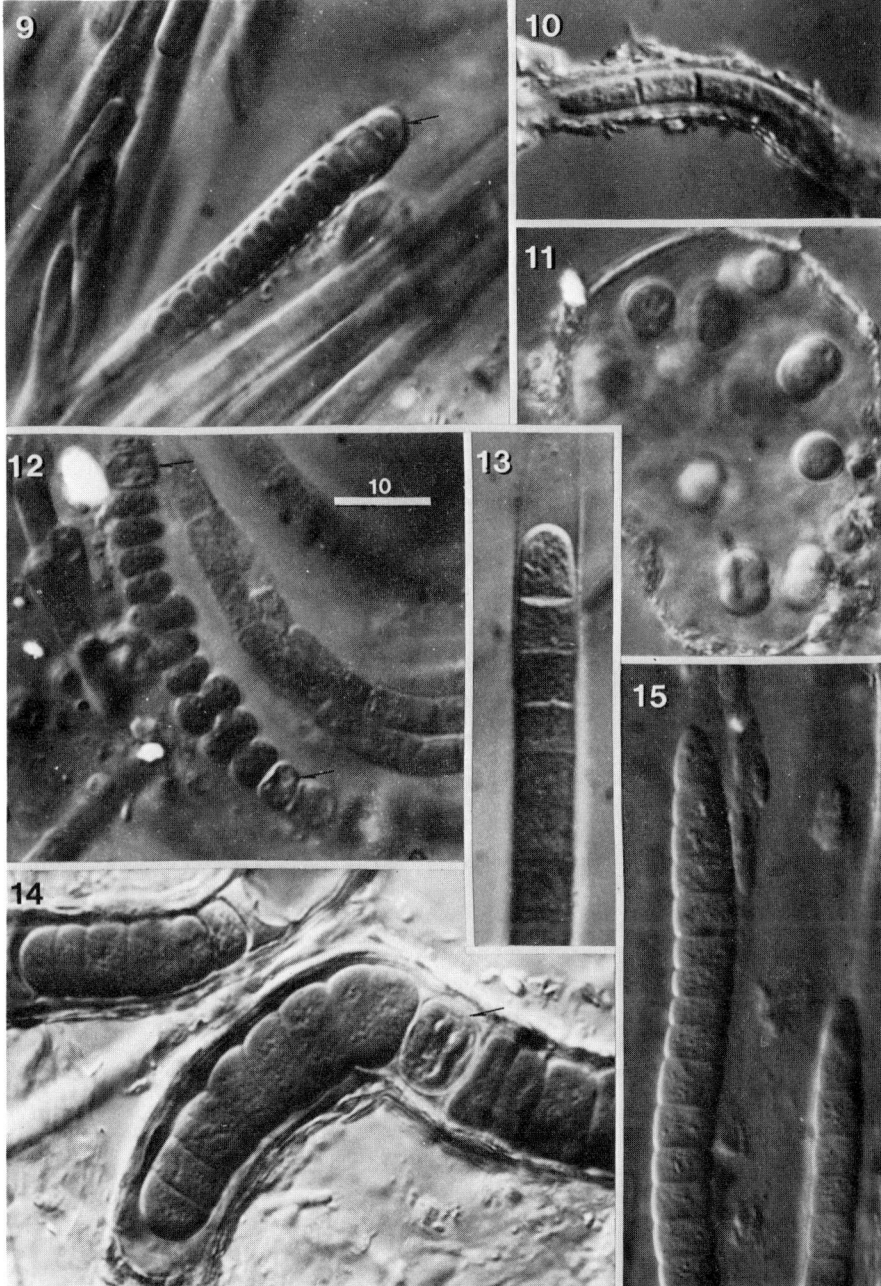

Fig. 9. *Calothrix pulvinata* Bornet-Thuret surrounded by trichomes of *Microcoleus chthonoplastes*. Heterocysts, sites of N_2 fixation, are marked with arrows in Figures 9, 12, and 14. Fig. 10. *Schizothrix penicillata* (Kützing) Gomont. Fig. 11. *Gloeocapsa nigrescens* (Vaucher) Novaček. Fig. 12. *Nostoc commune* Vaucher surrounded by trichomes of *M. chthonoplastes*. Fig. 13. *Phormidium retzii* Gomont. Fig. 14. *Scytonema tenellum* Gardner (has dark brown sheaths). Fig. 15. *Schizothrix subconstricta* Golubic.

Gomont (see Figure 7) and *M. Chthonoplastes* Thuret (see Figure 8), are the dominant organisms in both mats, and are responsible for approximately 90% of the organic material. *Microcoleus* is characterized by multiple trichomes that are bundled in a common sheath (see Figures 6, 7, and 8). Other species that are present in the Utah crust (which will be discussed for the remainder of this paper) are *Calothrix pulvinata* Bornet & Thuret (see Figure 9), *Schizothrix penicillata* (Kützing) Gomont (see Figure 10), *Nostoc commune* Vaucher (see Figure 12), *Gloeocapsa nigrescens* (Vaucher) Novaček (see Figure 11), *Phormidium retzii* Gomont (see Figure 13), *Scytonema tenellum* Gardner (see Figure 14), and *Schizothrix subconstricta* Golubic (see Figure 15). *Scytonema* and *Calothrix* are usually found at the surface of the mat while all others commonly occur beneath the surface.

The surface of the desert crust, when viewed with a dissecting microscope, consists of tightly packed sand, white sheaths abandoned by *Microcoleus*, dark brown sheaths of *Scytonema* and *Calothrix* which frequently contain trichomes, and small clumps of the foliose lichen *Collema*. When teased apart, the desert crust is seen to be composed of a dense mesh of filaments, many of which are dark green, to which grains of sand adhere. They are filaments of *Microcoleus*.

3.2. THE EFFECT OF WETTING AND EVAPORATION ON THE MAT

Fluctuation in salinity of the water retained in the desert soil takes place following every rain. First, a flushing of the mat with freshwater occurs, followed by a gradual increase in salinity due to solution of soil salts in interstitial water. They eventually form a brine, and, ultimately, precipitate as evaporation proceeds. A laboratory simulation of this process was conducted by placing an intact piece of mat in a finger bowl, and wetting it with distilled water. Within hours following wetting, a noticeable 'greening' of the newly glistening surface of the mat occurred, a phenomenon also described by Brock (1975). This 'greening' is actually due to the migration of *Microcoleus* trichomes from the interior of the mat to the surface. As the experiment progressed evaporation was retarded in the fingerbowl by covering it with glass. A community succession ensued, beginning with the appearance of leafy moss gametophytes. Soon after, a stalked sporophyte generation arose and fungal mycelia appeared. After several weeks of drying, the surface of the experimental mat was noticeably whiter than it had been before wetting. Patches of dark green were still visible, but these were mainly in depressions. Months later, the surface was almost entirely white except for the dry brown 'forest' of moss gametophytes. The white color was due to the empty white sheaths of *Microcoleus*. The smoother, rounded surface of the experimental mat is very hard and resistant to pressure compared with the brittle and fragile surface of desert crust collected in the field. Figure 4 shows a scanning electron micrograph of a vertical section through the hardened mat of the fingerbowl experiment. At the top, the bases of moss gametophytes rest on the horizontal layer which is composed mainly of strands of empty *Microcoleus* sheaths. When this mat was treated with dilute hydrochloric acid bubble formation occurred. Calcium carbonate had precipitated in the layer of abandoned algal sheaths (Figure 5); as a result of the prolonged wetness soil salts went into solution, they traveled toward the surface of the mat via capillary water transport as evaporation progressed and precipitated there.

3.3. THE RESPONSE OF THE CYANOPHYTES TO WETTING

The most significant biological property of desert crust is its ability to switch almost instantly from a dry dormant state to an active state. This is characterized by the striking motility of *Microcoleus* trichomes when wetted. Trichomes do not move when they are dry. Upon wetting, trichome motility occurs in two consecutive events which have different causes: (1) mechanical expulsion of trichomes due to swelling of the common sheath as it rapidly imbibes water; and (2) a self-propelled gliding motility of the trichomes. The process of trichome release from the common sheath is independent of the gliding motility of individual trichomes.

When a dry filament of *Microcoleus* is plucked from the mat and mounted dry for microscopy an immediate imbibition of water occurs as it is added, and the polysaccharide sheath swells to several times its original volume within seconds. The forces which result cause an explosive expulsion of the trichomes at rates of hundreds of microns per second.

Understanding of how the trichome propulsion occurs is based on observation of dehydrated, embedded filaments that have been thin sectioned and stained for light microscopy. A cross-section through a common sheath of *Microcoleus* shows that it is a composite of several smaller bundles, each visible as a dark ring. During dehydration the sheath pulls away from the trichomes it contains, leaving a space (Figure 6, arrow). Water enters such a tubular space by capillarity and the sheath expands in an inward as well as an outward direction during rehydration. This results in pressure on the trichomes which squeezes them out of the sheath. Usually, after a lag period of seconds to minutes, gliding motility of trichomes begins.

The mechanism of gliding motility is still a subject of debate. Cyanophyte trichomes have no visible locomotive structures and the theories that have been proposed were reviewed by Doetsch and Hageage (1968). *Microcoleus* trichomes move with a rotating motion forward or backward, frequently reversing direction.

There is no gliding of trichomes when desert crust is dehydrated; water, then, is one trigger for gliding motility. Figure 16 shows that light as well as water is a trigger for motility. It is a graph of the rate of gliding motility of the fastest *Microcoleus vaginatus* trichome (from a group of 50–100) that migrated on one illuminated agar plate for 12 hours before being placed in darkness for 12 more hours. It sometimes rains at night in the desert so we questioned whether light is necessary for trichome motility. Some trichomes that were plated in darkness and exposed to light only at the moment of observation were found to have migrated even in the absence of light, although the distance traveled was significantly shorter than in the daylight experiments. Figure 17 shows that water in the absence of light is a trigger for motility, although it is the initial

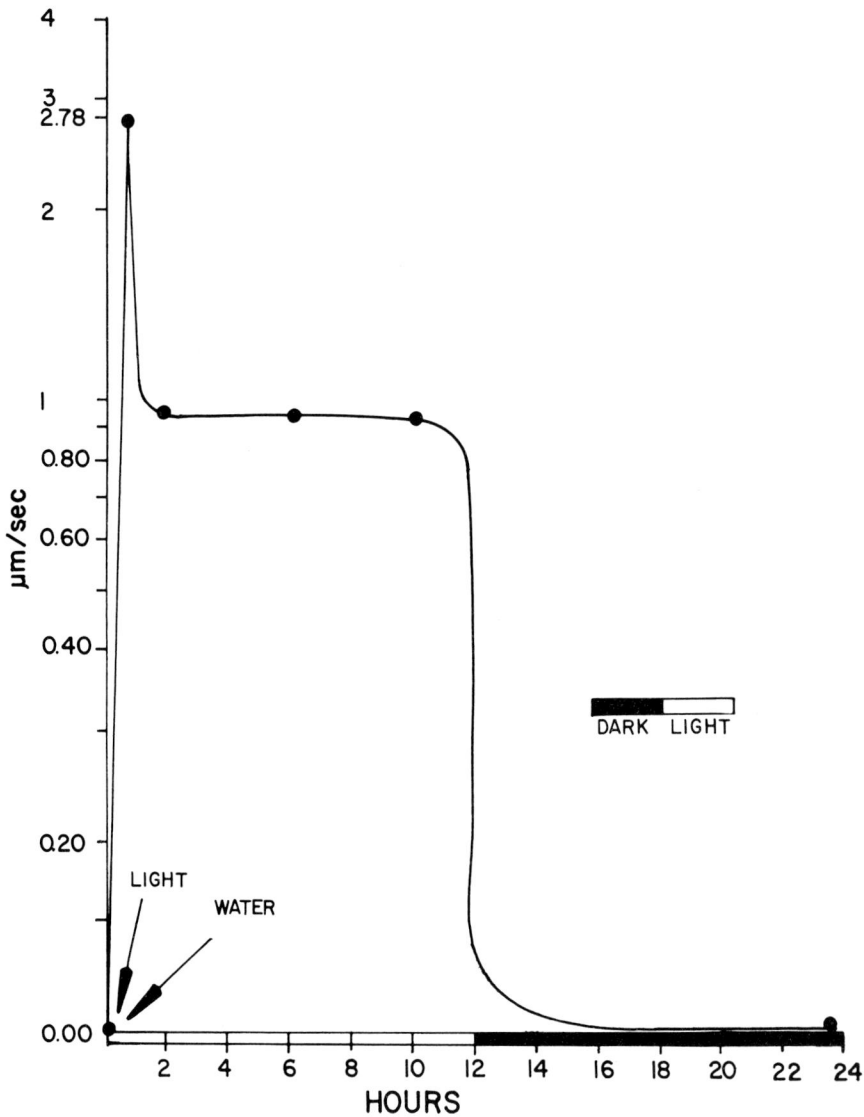

Fig. 16. Rates of gliding motility of a trichome of *M. vaginatus:* highest immediately following wetting on an illuminated agar plate. No motility occurred in the dark.

exposure to light that causes a significantly greater response. Following a 12 hour dark period in which group migration of trichomes had taken place, an average rate of 0.2 μm sec^{-1} was determined for the trichome in the lead. Then, this exemplary agar plate (as all others in the same test group) was exposed to light and observed continually. During the first 15 minutes there was a sharp increase in the leading trichome's rate of movement up to 11.0 μm sec^{-1}. Movement then slowed down to an average of 0.4 μm sec^{-1} until the end of the light period. This is twice the average rate for the dark period.

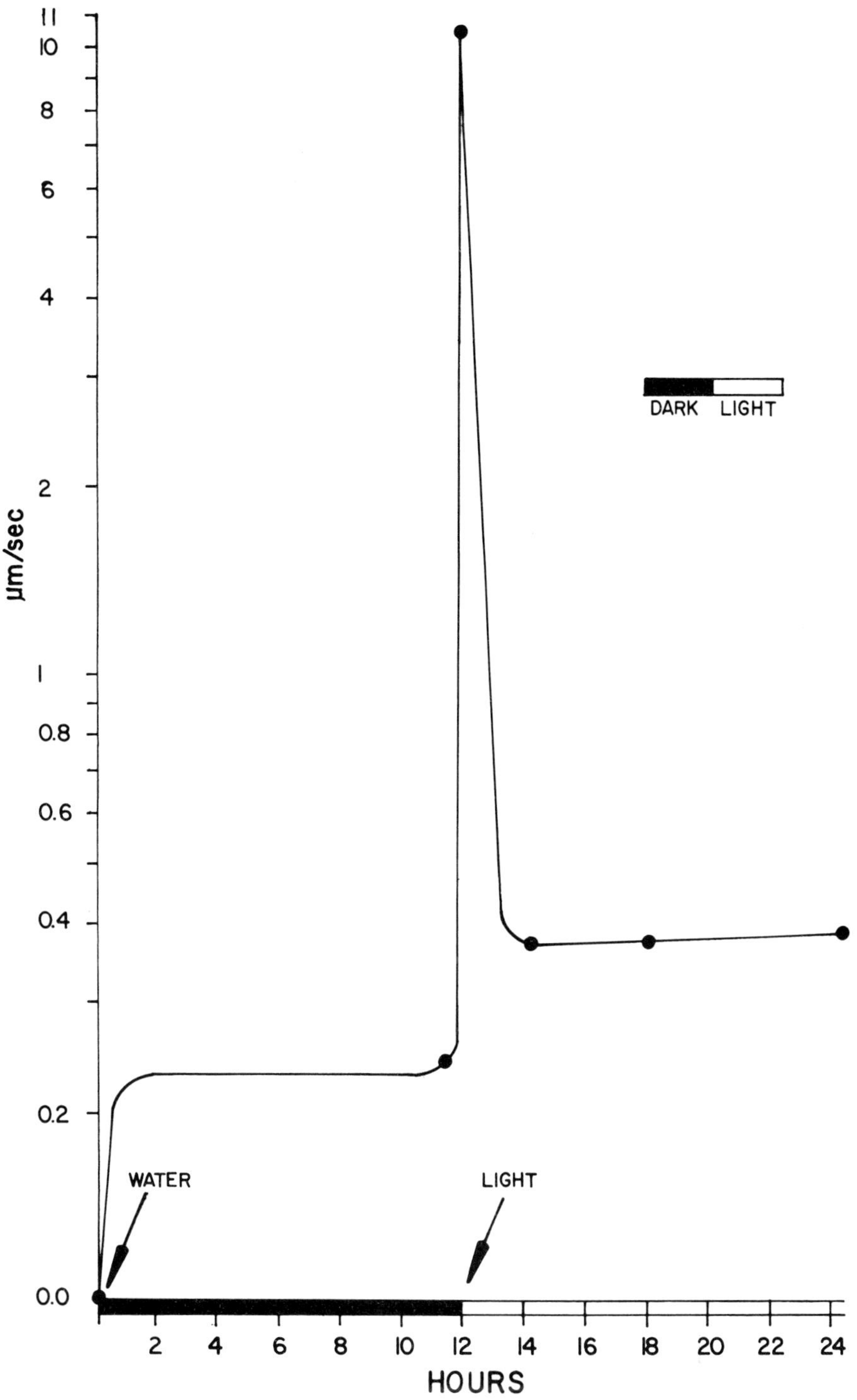

Fig. 17. Gliding motility of a trichome of *M. vaginatus* wetted in darkness on agar and subsequently exposed to light. Average speed was recorded for the dark period. The onset of light caused a burst of activity which then slowed down to a rate about twice that of the dark period.

The maximum distance traveled by a single trichome of *Microcoleus vaginatus* was 4.8 cm! Many trichomes are nearly 1 mm long, thus, such a trichome is capable of traveling about 500 times its own length on agar in one day. Assuming that slime production is continuous, the resulting 4.8 cm long thread is capable of trapping soil particles. Although it is not known whether rates of migration on agar are comparable to rates of migration on soil, they do provide a measure of potential for mat expansion. Desert crust formation and expansion does not depend directly on the growth and division rates of the organisms that make it, but rather, on their gliding motility and the accompanying production of a sheath which they continually abandon while migrating. It is the sheath which actually binds the mat, and represents the trails of single organisms which are capable of moving for days. Wind transported sand grains and clay particles adhere to the moist, sticky sheath and remain tightly bound to it as drying and sheath polymerization occurs. In this way soil accretes and is stabilized. The stabilized soil is very resistent to erosion by wind and water. Booth (1941) showed that when a similar mat and adjacent barren soil in Kansas were subjected to large amounts of forcefully applied water, the run-off from the barren soil was muddy while that of mat stabilized soil was quite clear.

3.4. THE RESPONSE OF THE CYANOPHYTES TO DRYING

The effect of dehydration on the metabolic activities of the *Microcoleus* dominated desert crust was studied by Brock (1975) who tested the mat's ability to photosynthesize under conditions of reduced matric and osmotic water potential by measuring the production of chlorophyll. Brock dehydrated entire pieces of mat either by absorption (reduction of matric water potential), or by wetting it with water of increasing salinity (reduction of osmotic water potential). Matric reduction caused a precipitous decrease in production of chlorophyll by the organisms at -10 bars, while osmotic reduction of water potential caused an *increase* in chlorophyll production from 0 to -20 bars, and thereafter, at lower osmotic water potential values, chlorophyll production decreased markedly. Brock calculated the water potential of sea water to be -28 bars and concluded that desert crust is not active at marine salinities. However, according to conversion tables (Hale, 1965) a salinity of 30–35°/oo (average salinity of the ocean) corresponds to a water potential range of -20 to -25 bars. The photosynthetic activity of the mat that Brock measured was high within this range. It is interesting to note that while *Microcoleus vaginatus, Schizothrix penicillata, Gloeocapsa nigrescens, Nostoc commune*, and *Scytonema tenellum* are freshwater cyanophytes, other members of the mat community (e.g. *Microcoleus chthonoplastes, Schizothrix subconstricta*, and *Calothrix pulvinata*) commonly occur in marine environments. In fact, the Sippewisset Marsh intertidal algal mats (Wood Hole, Massachusetts) contain a large proportion of *Microcoleus chthonoplastes*. This species is also present in benthic mats of Solar Lake (Israel) near the surface of the pond (Campbell, unpublished observations) which undergoes seasonal fluctuation in salinity that culminates in precipitation of salt at the boundaries of the pond (Cohen et al., 1977). Thus, the species composition of the Utah desert crust makes it well adapted to fluctuations in salinity, especially because the two dominating species may have different salinity optima (*Microcoleus vaginatus* vs. *M. chthonoplastes*). However, both species react to reduced matric water potential by dehydration and sudden collapse of the sheath with an accompanying loss of metabolic activity.

The ultrastructure of the cyanophyte sheath is a finely meshed network of thin fibrils (Fogg et al., 1973). In desert crust, as dehydration occurs, this polysaccharide network collapses on itself, and, as a result, it increases in density which serves to slow additional water loss. Due to its small size, changes in the sheath morphology of *Microcoleus* are difficult to observe. However, observations of a marine *Rivularia* sp. during alcohol dehydration showed that from 0 to 95% ethanol no perceptible alteration of the size or shape of sheaths occurred. Then, as the ethanol concentration increased, the sheath suddenly collapsed to about 75% of its original size (Campbell and Golubic, unpublished observations). This indicates that strong molecular forces in the sheath bind water and maintain the sheath in expanded form until a critical amount of water is removed. If a similar affinity for water is characteristic of the *Microcoleus* sheath, the effective result would be hydrated surroundings for the trichomes up to the moment that the sheath collapse occurred. Such a mechanism could be instrumental in allowing sufficient time for the trichomes to slow or switch off their metabolic activity as dehydration ensues, thereby modifying the harsh desert climatic extremes.

The sudden and rapid changes in availability of water in the desert preclude colonization of soil by cyanophytes that are exclusively aquatic. Only those cyanophytes which produce sheaths can survive rapid desiccation and are competitive. Except for *Gloeocapsa*, all others in the assemblage are motile, either as fully differentiated trichomes or in their juvenile hormogonal stages. Resumption of metabolic activity upon rehydration is evidenced by the immediate migration of trichomes (following mechanical expulsion from a common sheath) and the accompanying renewed production of hygroscopic sheath. Rapid resumption of metabolic activity, gliding motility, and tolerance to drying are adaptations which allow the formation and spreading of vast communities of the prokaryotic primary producers that blanket otherwise barren terrain in an environment which fluctuates between the inundation of flash floods and long periods of drought.

3.5. THE FOSSILIZATION POTENTIAL OF DESERT CRUST

Desert crust does not normally lithify. It is a fragile structure, and a transient phenomenon which is easily trampled. However, the fingerbowl experiment showed that carbonate may precipitate in the mat. Similar conditions of prolonged wetting and then drying in nature could allow formation of caliche. If this process were repeated many times a carbonate rock could result, similar to that described by Swineford et al. (1958) who proposed a subaerial caliche development for limestone formations of the Great

Plains. Sufficient fossilized microflora had been previously found at the same site by Elias (1931) that he termed it "algal limestone". Swineford *et al.* denied finding microfossils and also rejected the lacustrine environment of deposition that had been postulated by Elias, arguing that the time of deposition occurred at the close of the Pliocene "at the culmination of a time trending toward aridity in the Great Plains". They noted shrinkage cracks, barite and calcite veins, and rotated blocks which they said were attributable to "successive episodes of desiccation, shrinkage, solubilization and reprecipitation". Although the limestone studied by Swineford *et al.* need not have required the presence of an algal mat such as desert crust for its formation, it is important to note that the presence of such a mat could be responsible for, or indicative of calcrete formation under similar conditions. James (1972) and Krumbein (1968) discuss the possible involvement of microorganisms in the production of some calcrete carbonates. Krumbein and Lange (in press) show direct involvement of blue-green algae in the formation of "desert stromatolites" of Israel. These observations indicate that even a fragile terrestrial desert crust community has a preservation potential and may appear in fossil calcrete deposits as lithified microscopic tubes (sheaths) and cellular remains of algal trichomes. Thus, a reexamination of calcrete limestone previously identified as resulting from abiogenic subsoil caliche formation is warranted.

3.6. PRECAMBRIAN SOILS

The colonization of land in the early Precambrian by prokaryotic, lithophytic assemblages similar to those that presently occur on alpine cliffs was suggested by Golubic and Campbell (1979). This proposal was based on the morphological similarity of *Eosynechococcus moorei*, a marine intertidal cyanophyte in the Precambrian Belcher Island Formation, to *Gloeothece coerulea*, a modern freshwater lithophyte.

Schopf (1968) suggested that "Oscillatorian gliding" may have enabled early colonization of land surfaces. The present study has established the adaptational value of gliding motility and the accompanying sheath production in the modern oscillatoriacean cyanophyte *Microcoleus* for colonization of barren land and accretion of soil. The antiquity of *Microcoleus vaginatus*, one of the desert crust species, has been invoked by Schopf's interpretation of the Late Precambrian *Cephalophytarion grande* (Schopf, 1968).

Several cyanophytes show extreme morphological conservatism, i.e. they have maintained species level similarity over time spans of billions of years (Knoll *et al.*, 1975; Golubic and Hofmann, 1976; Golubic and Campbell, 1979). It is important to note that some of these fossil-Recent counterparts, *e.g. Palaeolyngbya barghoorniana-Lyngbya estuarii* and *Eoentophysalis belcherensis-Entophysalis major*, have continued to inhabit marine intertidal environments since the Precambrian, while others occurred in the intertidal zone of Precambrian seas but have terrestrial (or freshwater) modern equivalents. The intertidal *Cephalophytarion grande* and the soil dwelling *Microcoleus vaginatus* represent another such case, illustrating that the evolutionary transition of cyanophytes from the marine environment to freshwater-wetted land was initially achieved by organisms that had occupied intertidal niches where they were exposed to a daily alternation of wetting and drying, and wide ranges of salinity. Such organisms would have been preadapted for invasion of dry land. In desert crust, one co-dominant species is *Microcoleus vaginatus*, a freshwater alga, and the other, *M. chthonoplastes*, is marine. Both are capable of tolerating the wide fluctuation in salinity that occurs in the wetted desert soil, although they probably have different optima. With respect to salinity fluctuation, dehydration, and rehydration, the desert soil is comparable to the intertidal sands.

Fisher (1965), Roscoe (1969), Tappan and Loeblich (1971), Kalliokoski (1975), and Fryer (1977) have noted or discussed the occurrence of mature Precambrian paleosols. D. Grandstaff (personal communication, 1979) has recorded the presence of 0.25% reduced organic carbon in analyses of a 15m thick, subaerially formed paleosol from the Precambrian (2.4 billion years old) Blind River Formation of Ontario. What was the source of organic matter in the subaerially formed soil? It is now proposed that terrestrial blue-green algal mats may have been responsible for the accumulation, stabilization, and biogenic modification of mature Precambrian soils. In its ecological function the modern *Microcoleus* dominated desert algal mat provides a representative model for such an ancient subaerial microbial community.

Acknowledgments

Special thanks are due to P. and D. Johnson, rangers in Canyonlands National Park who hosted me during the field work, provided the photographs for Figures 1-3, and sent additional samples of desert crust for laboratory studies in Boston. B. Doran assisted in field collections and critically read the manuscript. J. Waterbury gave valuable advice for the fixation and resin embedding of the cyanophytes. E. Selling of MCZ Harvard, provided expert assistance at the scanning electron microscope. B. Cameron, G. Cooper-Driver, E. Hoffman, W. Krumbein, and L. Margulis are thanked for critical reading of the manuscript. I am especially grateful to S. Golubic for his advice and criticism throughout the project which is submitted in partial fulfillment of requirements for a Ph.D. degree at Boston University. Research was supported by NSF grants GA43391 to S. Golubic, EAR 76-84233, EAR 76-84233 A01 to S. Golubic and B. Cameron, and NASA grant NSG-7588 to S. Golubic.

References

Booth, W. E.: 1941, *Ecology* 22, 38.
Brock, T. D.: 1975, *J. Phycol.* 11, 316.
Cohen, Y., Krumbein, W. E., Goldberg, M., and Shilo, M.: 1977, *Limnol. and Ocean.* 22, 597.
Doetsch, R. N. and Hageage, G. J.: 1968, *Biol. Rev.* 43, 317.
Elias, M. K.: 1931, *Kansas Geol. Survey Bull.* 18, 1.
Fisher, A. G.: 1965, *Proc. Nat. Acad. Sci.* 53, 1205.
Fogg, G. E., Stewart, W. D. P., Fay, P., and Walsby, A. E.: 1973, *The blue-green algae*, Academic Press, London and New York, p. 74.
Fryer, B. J., 1977. *Geol. Abstr. and Proc. of 25th Ann. Meeting of the Lake Superior Conference.*

Golubic, S.: 1976, in M. R. Walter (ed.), *Stromatolites,* Developments in Sedimentology, 20, Elsevier, Amsterdam, Oxford and New York, p. 113.

Golubic, S. and Hofmann, H. J.: 1976, *J. Paleontol.* **50**, 1074.

Golubic, S. and Campbell, S.: 1979, *Precambrian Research* **8**, 201.

Hale, L. J.: 1965. *Biological Laboratory Data,* Pitman Press, Bath, p. 108.

James, N. P.: 1972, *J. Sediment. Petrol.* **42**, 817.

Kalliokoski, J.: 1975, *G.S.A. Bull.,* **86**, 371.

Knoll, A. H., Barghoorn, E. S. and Golubic, S.: 1975, *Proc. Nat. Acad. Sci.* **72**, 2488.

Krumbein, W. E.: 1968, in G. Muller and G. M. Friedman (eds.) *Recent developments in carbonate sedimentology in central Europe*, Springer Verlag, New York p. 138.

Krumbein, W. E. and Lange, C.: *Sedimentology*, in press.

Rippka, R., Waterbury, J. and Cohen-Bazire, G.: 1974. *Arch. Microbiol.* **11**, 419.

Roscoe, S. M.: 1969, *G. S. Canada Paper* **68-40**.

Schopf, J. W.: 1968, *J. Paleontol.* **42**, 651.

Shields, L. M., Mitchell, C. and Drouet, F.: 1957, *Am. J. Bot.* **44**, 489.

Stanier, R. Y., Kunisawa, R., Mandel, M., Cohen-Bazire, G.: 1971, *Bact. Rev.* **35**, 171.

Swineford, A., Leonard, A. B., and Frye, J. C.: 1958. *State Geological Survey of Kansas, Bull.* **130**, 97.

Tappan, H. and Loeblich, A. R.: 1971, in R. Kosanke and A. T. Cross (eds.) *Symposium on palynology of the Late Cretaceous and Early Tertiary*, Geol. Soc. Amer. Spec. Paper **127**, 247.

THE EVOLUTION OF THE SOLAR 'CONSTANT'

MICHAEL J. NEWMAN*

Max Planck Institut für Physik und Astrophysik, Föhringer Ring 6, D-8000 München 40, F.R.G.

Abstract. The ultimate source of the energy utilized by life on Earth is the Sun, and the behavior of the Sun determines to a large extent the conditions under which life originated and continues to thrive. What can we say about the history of the Sun? Has the solar 'constant', the rate at which energy is received by the Earth from the Sun per unit area per unit time, been constant at its present level since Archean times? Three mechanisms by which it has been suggested that the solar energy output can vary with time will be discussed, characterized by long ($\sim 10^9$ yr), intermediate ($\sim 10^8$ yr), and short ($\sim$ years to decades) time scales.

1. Introduction

Although the rate at which energy is supplied to the Earth by the Sun is generally regarded as being constant in time, there are sound astrophysical reasons for expecting a time variation in this important quantity beyond the well-known modulation of the solar 'constant' due to changes in the details of the Earth's orbit. Three possible types of variation will be discussed:

(1) The inexorable increase of the solar luminosity over geological timescales as the conversion of hydrogen into helium, which provides the thermonuclear power of the Sun, slowly increases the mean molecular weight in its interior. This luminosity increase, amounting to some 25% over the 4.7×10^9 yr lifetime of the Sun, has been shown to be a nearly model-independent result not affected by the uncertainties arising from the difficulties concerning the interpretation of the solar neutrino experiment.

(2) The infrequent temporary enhancement of the solar luminosity due to the gravitational energy release of material accreted onto the solar surface as the solar system traverses a dense interstellar cloud. This luminosity perturbation, which may have occurred half a dozen or more times during the Sun's history, is peaked in the short-wavelength region of the spectrum, and may constitute a serious hazard for life on Earth.

(3) The small amplitude rapid fluctuations of the solar luminosity which may arise due to the finite efficiency and stochastic nature of convective energy transport.

The latter effect, while it should be less important for the evolution of life on our planet than the others, differs in being more accessible to direct experimental measurement than the long-timescale effects, and may have a significant impact on climatic change in the current epoch.

---

* Present address: Theoretical Division, Los Alamos Scientific Laboratory, University of California, Los Alamos, NM 87545, U.S.A.

2. Accumulation of He Ash

Most detailed numerical models of the structure and evolution of the Sun show a roughly 25% luminosity increase from the time of thermonuclear ignition to the present, and Newman and Rood (1977) have noted that this behavior is a direct consequence of energy generation by consumption of a light nuclear fuel (hydrogen) and accumulation of waste products (helium). It is therefore very difficult to avoid. A simple dimensional analysis of the differential equations of stellar structure (Cox and Giuli, 1968) indicates that the luminosity should scale as roughly

$$L \propto \frac{ac}{\kappa_0} M^{5.5} R^{-0.5} \left(\frac{G m_p \mu}{k}\right)^{7.5}, \tag{1}$$

where M is the solar mass, R the solar radius, a is the radiation pressure constant $4\sigma/c$, κ_0 is the opacity coefficient, G is the gravitational constant, m_p is the proton mass, k is Boltzman's constant, and μ is the mean molecular weight (Clayton, 1968). If the solar mass and the fundamental constants remain constant then the logarithmic time derivative of the solar luminosity should be approximately

$$\frac{1}{L}\frac{dL}{dt} \approx -0.5 \frac{1}{R}\frac{dR}{dt} + 7.5 \frac{1}{\mu}\frac{d\mu}{dt}. \tag{2}$$

Thus if the source of solar energy is to a small extent gravitational contraction ($dR/dt < 0$) and principally the conversion of hydrogen into helium ($d\mu/dt > 0$), then the solar luminosity must be an increasing function of time. A quantitative estimate of the rate of change can even be made by this simple argument. If the solar luminosity is provided by the release of $\epsilon = 6.4 \times 10^{18}$ ergs per gram of hydrogen converted into helium, then the average hydrogen mass fraction X in the solar interior must be changing at a rate

$$\frac{dX}{dt} \approx -\frac{L}{M\epsilon}$$

$$\approx -0.01/10^9 \text{ yr}, \tag{3}$$

and from

$$\mu \approx \frac{\frac{4}{3}}{1 + \frac{5}{3} X} \tag{4}$$

we have immediately from Equation (2)

$$\frac{1}{L}\frac{dL}{dt} \approx \frac{L}{M\epsilon} \frac{12.5}{1 + \frac{5}{3} X} \tag{5}$$

$$\approx 0.05/10^9 \text{ yr}$$

if the Sun is composed primarily of hydrogen. As Newman and Rood showed, the luminosity enhancement predicted by (5) is shared by many classes of solar models, including exotic models constructed to be in agreement with the Brookhaven solar neutrino experiment (Davis, 1978). Even these models share the basic assumption that the Sun is powered by thermonuclear reactions, and the increase in mean molecular weight resulting from fuel exhaustion and accumulation of nuclear 'ashes' directly results in an increasing solar luminosity. As we see from Equation (1), almost the only way to avoid the luminosity increase of some 5 to 7% per 10^9 yr is to appeal to changes in the solar mass, radius, or one of the fundamental physical constants. The present solar wind is not sufficient to cause significant mass loss, and *increasing* the solar mass, as by accretion, only serves to further enhance the luminosity increase. Similarly the solar radius is thought to be very slowly *decreasing* under the influence of gravity, contributing slightly to the luminosity increase. It is difficult to increase the radius unless the rate of energy generation in the interior is increased — and we have seen that that is the major source of the luminosity enhancement. The predicted slow-but-steady increase of the solar luminosity over cosmic time scales would thus seem to be on very firm ground, unless the physical constants do not remain constant in time.

3. Cosmic Pollution

McCrea (1975) has recalled the suggestion of Hoyle and Lyttleton (1939) that the accretion of matter from the interstellar medium by the Sun acts as a trigger for the occurrence of Ice Ages on the Earth, through a complicated mechanism in which the luminosity perturbation resulting from the gravitational energy release of the matter falling onto the solar surface increases the rate of evaporation, which increases the cloud cover and the rate of precipitation, increasing the albedo of the planet and leading to glaciation. However, the climatic response of the Earth involves so many complex feedback links that it is not certain that even the algebraic sign of the global response to a perturbation of the solar luminosity is understood. Nonetheless, it seems clear that passage of the solar system through a dense interstellar cloud would have some impact on the terrestrial climate, whether or not we can state with confidence what the effect would be. Possible climatic consequences of such an encounter have been discussed by Begelman and Rees (1976) and Talbot *et al.* (1977). Encounters with clouds of sufficiently high density and low relative velocity to yield the high levels of accretion discussed by Hoyle and Lyttleton or McCrea would seem to be rare (less than one such encounter expected during the lifetime of the solar system — which is just as well, since Talbot *et al.* have estimated that the resulting uv flux could be deadly for life on Earth. However, encounters with more modest clouds should occur many times during the lifetime of a star like our Sun, and encounters sufficient to cause significant climatic effects on Earth may occur at intervals averaging of order 10^8 yr. Talbot and Newman would expect the occurrence of such events to be distributed randomly in time, reflecting the stochastic nature of collisions between randomly distributed objects, but McCrea has suggested that passage of the solar system through dense interstellar clouds occurs preferentially during the crossing of the spiral arms of our galaxy, which occurs at intervals of roughly 250 million years, and that there is a correlation between the time scales describing the motion of the solar system in the galaxy and the occurrence of epochs of glaciation in the paleoclimate of the Earth. Whether the occurrence is as regular as McCrea suggests, and whether the climatic impact is through the luminosity perturbation or through other mechanisms which have been suggested — such as compression of the solar wind cavity, dust loading of the Earth's atmosphere, or even the influence of supernova explosions which occur preferentially in or near dense clouds where massive young stars are thought to form — it seems clear that encounters between the solar system and dense interstellar clouds can have a significant impact on conditions for life on Earth. Whether or not they have actually done so in the past, and how often, is currently a matter of controversy.

4. Stochastic Convection

While the Sun's energy is thought to be carried outward from its site of production (by thermonuclear reactions deep in its interior) primarily by radiative transport processes throughout most of its mass, there is a relatively thin outer region where energy transport is primarily by convective motion. The quantitative understanding of energy transport by convection is poor, and most astrophysical calculations are done with the mixing length theory of convection. This assumes that a typical fluid element moves on the average a distance equal to the mixing length l before it mixes with its surroundings and gives up its excess heat energy. It is further assumed that the ratio $\alpha = l/H_p$ of the mixing length to the pressure scale height H_p (the distance over which the pressure changes by a factor of e, the base of natural logarithms) is a constant, usually taken to be about 1.5. Solar models, in particular, are 'tuned' by adjusting the mixing length parameter α until the model radius is equal to the observed solar radius (the radius of a stellar model is a sensitive function of α). Dearborn and Newman (1978), however, have questioned how precisely we can characterize a complex stochastic process like convection by a single average quantity like α, and consider the consequences if the appropriate average value of α changes with time. If α is increased, convection becomes more efficient, and the same flux can be carried with a smaller temperature gradient. Then the convective envelope contracts slightly. If α is decreased, convection becomes less efficient and a larger temperature gradient is required to produce the same net energy transport. Then the envelope expands slightly. In either case there is an interchange between the gravitational potential energy and internal energy of the material in the convection zone, and a luminosity perturbation results. The change in luminosity can be estimated as

$$\delta L \approx \frac{GM\Delta m}{R^2} \frac{H_p \delta \alpha}{\tau}, \qquad (6)$$

where we have expressed the change in gravitational energy in terms of the change in the mixing length parameter, Δm is the mass of the convection zone, and τ is the timescale on which the energy is released or absorbed. If the timescale τ_α on which α is changing is longer than the thermal response time of the envelope $\tau_c \sim 10^5$ yr then $\tau = \tau_\alpha$ and the luminosity perturbation depends on the rate of change of the mixing length parameter as found by Ulrich (1975). However, if α is changing on time scales which are short compared to the thermal timescale of the envelope ($\tau_\alpha \ll \tau_c$) then, although the structure of the envelope can adjust on the dynamic timescale ~ 10 m, the excess energy can be radiated away or absorbed only on the timescale $\tau = \tau_c$, and the luminosity perturbation depends only on the amplitude $\delta\alpha$. Evaluating Equation (6) numerically for a standard solar model yields

$$\delta \log L \approx 0.2 \delta\alpha, \tag{7}$$

which has been confirmed to good accuracy with detailed numerical models. The luminosity perturbation given by (7) would decay away on the timescale τ_c if no new perturbation $\delta\alpha$ is introduced; if changes in α are frequent the luminosity will track with the response function (7). The fine structure of the solar luminosity could thus be quite jagged, depending on the behavior of α. The solar constant is not monitored to much better than 1%, and 1% changes in L would result from fluctuations as small as $\delta\alpha \approx 0.02$. Since significant climatic effects could result from luminosity excursions at this level, it is a matter of concern that our current understanding of convection cannot rule them out. Dearborn and Newman mentioned the influence of magnetic fields on convection, the limited number of supergranules (which may be directly related to the convective cells themselves) observed on the surface of the Sun, and the inherently random nature of convective motions as reasons for suspecting fluctuations in the efficiency of convection may exist. Whether such effects are responsible for the apparent solar variations reported by Livingston *et al.* (1977) and White and Livingston (1978) is not yet known. The luminosity fluctuations discussed here would be manifested on the time scale on which the efficiency of convection is changing, which is most likely months, years, or decades depending on the mechanism involved, although longer time scales have also been discussed. They may thus be responsible for climatic effects in the present epoch, although it is not yet certain whether they occur.

5. Conclusions

Three mechanisms have been discussed by which the solar luminosity may change in time. The slow steady increase in solar luminosity of some 5% per 10^9 yr is a fundamental consequence of energy generation by thermonuclear reactions, and is difficult to avoid unless fundamental physical constants do not remain constant over cosmic timescales. Very large ($\gtrsim 10\%$) luminosity excursions due to encounters between the solar system and interstellar clouds of high density are unlikely to have occurred, but cannot be ruled out. Their effect on conditions for life would be profound. More modest ($\lesssim 1\%$) luminosity enhancements due to encounters with clouds of lower density or larger relative velocity may have occurred half a dozen or more times during the history of the solar system, and could have devastating effects on life, since the luminosity perturbation due to solar accretion of interstellar matter is peaked at short wavelengths. Encounters with clouds of modest density producing small ($\lesssim 0.1\%$) luminosity enhancements may have occurred 50 or more times since the formation of the solar system, and could produce significant climatic effects through a variety of mechanisms, not all of them connected with the resulting distortion of the solar spectrum. These effects are on less certain ground than the long-time-scale enhancement due to fuel consumption, but encounters between the solar system and dense interstellar clouds should have had some impact on the conditions under which life has developed on Earth. Finally, it has been suggested that fluctuations of the efficiency of convection result in small amplitude variations of the solar luminosity on relatively short time scales. These have not been shown to occur, but may be important for the fine structure of the paleotemperature curves if they are real.

It is not likely that the solar 'constant' has remained strictly constant, and time variation of this important quantity may have had significant impact on the development of life on our planet.

Acknowledgment

Los Alamos Scientific Laboratory is operated under the auspices of the United States Department of Energy.

References

Begelman, M. C. and Rees, M. J.: 1976, *Nature* **261**, 298.
Clayton, D. D.: 1968, *Principles of Stellar Evolution and Nucleosynthesis*, McGraw-Hill, New York.
Cox, J. P. and Giuli, R. T.: 1968, *Principles of Stellar Structure*, Gordon and Breach, New York.
Davis, R., Jr.: 1978, in G. Friedlander (ed.), *Proceedings of the Informal Conference on Status and Future of Solar Neutrino Research*, Brookhaven National Laboratory, Jan. 5–7, 1978, BNL 50879, Vol. 1, p. 1.
Dearborn, D. S. P. and Newman, M. J.: 1978, *Science* **201**, 150.
Hoyle, F. and Lyttleton, R. A.: 1939, *Proc. Cambridge Phil. Soc.* **25**, 405.
Livingston, W., Milkey, R., and Slaughter, C.: 1977, *Astrophys. J.* **211**, 281.
McCrea, W. H.: 1975, *Nature* **255**, 607.
Newman, M. J. and Rood, R. T.: 1977, *Science* **198**, 1035.
Talbot, R. J., Jr. and Newman, M. J.: 1977, *Astrophys. J. Suppl.* **34**, 295.
Talbot, R. J., Jr., Butler, D. M., and Newman, M. J.: 1976, *Nature* **262**, 561.
Ulrich, R. K.: 1975, *Science* **190**, 619.
White, O. R. and Livingston, W.: 1978, *Astrophys. J.* **226**, 679.

OZONE, ULTRAVIOLET FLUX AND TEMPERATURE OF THE PALEOATMOSPHERE

JOEL S. LEVINE, ROBERT E. BOUGHNER, and KATHRYN A. SMITH

NASA Langley Research Center, Hampton, VA 23665, U.S.A.

Abstract. Of all tropospheric species, ozone (O_3) comes closest to being naturally present at toxic levels. In addition, O_3 controls the ultraviolet flux reaching the Earth's surface and affects the temperature of the surface and atmosphere. For these reasons, O_3 was an important species of the paleoatmosphere. Surface and atmospheric levels of paleoatmospheric O_3 were calculated using a detailed photochemical model, including the chemistry of the oxygen, nitrogen, and hydrogen species and the effects of vertical transport. Surface and tropospheric O_3, as well as the total O_3 column, were found to maximize for an atmospheric oxygen level of 10^{-1} present atmospheric level (PAL). Coupled photochemical/radiative-convective calculations indicate that the radiative effects of O_3 corresponding to an oxygen level of 10^{-1} PAL resulted in a globally-averaged surface temperature increase of 4.5 K.

1. Introduction

Of major importance to the understanding of the origin and evolution of life is knowledge about the parameters that determined the environmental conditions suitable for the emergence and continuation of life. Such a parameter was the concentration of ozone (O_3) in the paleoatmosphere. O_3 controlled the ultraviolet flux reaching the Earth's surface and affected the temperature of both the surface and the atmosphere. In addition, of all tropospheric species, O_3 comes closest to being naturally present at toxic levels (i.e., see papers on tropospheric chemistry, Levine and Schryer (eds.), 1978).

Life on the surface of planet Earth is protected from lethal solar ultraviolet radiation between 200 and 300 nm by a thin layer of O_3 concentrated at a height of about 25 km. Over geological time, the amount of solar ultraviolet radiation reaching the surface has been controlled by the level of atmospheric O_3, which has varied considerably. To study the variation of surface ultraviolet radiation, we have investigated the evolution of O_3 in the Earth's atmosphere using a detailed photochemical model (Levine *et al.*, 1979).

The appearance and evolution of O_3 was strongly coupled to the appearance and evolution of molecular oxygen (O_2). The first investigation of the evolution of O_3 in the O_2-deficient paleoatmosphere was the qualitative treatment of Berkner and Marshall (1965). Next, Ratner and Walker (1972) used a simple photochemical model – the four Chapman reactions for a pure O_2 atmosphere, without transport – to investigate the evolution of O_3. Hesstvedt *et al.* (1974) added the hydrogen species chemistry to the Chapman reactions in their study. Blake and Carver (1977) added the nitrogen species chemistry with the exception of nitrous oxide to the hydrogen and oxygen species chemistry in a study of the evolution of O_3 and assumed photochemical equilibrium; i.e., they did not include the effect of vertical transport on species distribution in their calculations. In a more recent study of the evolution of O_3, Levine et al. (1979) included the chemistry of the oxygen, hydrogen, and nitrogen species, plus the effect of vertical transport on the distribution of the calculated atmospheric species. This study was the first to include nitrous oxide (N_2O) which is produced via denitrification by soil bacteria (Bates and Hays, 1967; Crutzen, 1970; McElroy and McConnell, 1971). The oxidation of N_2O is the major source of the nitrogen oxides which control O_3 levels in the present stratosphere (Crutzen, 1970). In addition, Levine et al. (1979) discussed the nitrogen and hydrogen species concentrations of the O_2-deficient paleoatmosphere and the variation of these species as O_2 evolved to present atmospheric levels. Their assumptions and results are summarized in Sections 2 and 3, respectively. The O_3 profiles discussed in Section 3 were used to calculate the levels of surface ultraviolet flux discussed in Section 4 and for the surface and atmospheric temperature calculations discussed in Section 5.

Our knowledge about atmospheric reactions and reaction rates in general, and the photochemistry of stratospheric O_3, in particular, increased at a very rapid rate during the mid and late 1970's, due primarily to concerns about the inadvertent depletion of stratospheric O_3 by anthropogenic activities, e.g., supersonic transports, fluorocarbons — used as propellants in aerosol spray cans — and increased useage of nitrogen agricultural fertilizer. Our new understanding of the photochemistry of stratospheric O_3 was evident by the more complex and sophisticated photochemical models developed to study the evolution of O_3 over geological time. In the present atmosphere, the bulk of atmospheric O_3 is found in the stratosphere. To-date, photochemical models used to study the evolution of O_3 have been primarily stratospheric photochemical models, e.g., Levine et al. (1979). Future investigations of the evolution of O_3 on our planet should also include relevant tropospheric photochemical and chemical processes. Relevant tropospheric chemical processes include the methane oxidation chain, the rainout of soluble tropospheric trace species, e.g., HNO_3 and H_2O_2, realistic surface fluxes of biogenic trace species, e.g., N_2O and CH_4 and the inclusion of the surface as a physical sink for O_3. In addition, future studies of the evolution of O_3 should consider the chemistry of carbon and chlorine species, in addition to the oxygen, hydrogen and nitrogen species. Chlorine species resulting from sea salt particles and from volcanic activity may have played an important role in evolution of O_3 in the O_2-deficient paleoatmosphere.

2. Composition and Structure of the Paleoatmosphere

One uncertainty in our study of the evolution of O_3 as a function of evolving O_2 level concerns the composition and structure of the paleoatmosphere during the period that O_2 rose from 10^{-4} of its present atmospheric level (10^{-4} PAL) to its present atmospheric level (1 PAL). This uncertainty results in part from our lack of knowledge concerning the exact chronology for the evolution of O_2. For example, Berkner and Marshall (1965) have speculated that O_2 rose from 10^{-3} PAL to its present level in the more recent past, over the last 600 million years, whereas Walker (1977b) has suggested that O_2 "rose rapidly from essentially zero to approximately its present value (within a factor of 10) ... probably about 2 billion years ago". For our photochemical calculations we need to know the approximate concentrations of nitrogen (N_2), water vapor (H_2O), carbon dioxide (CO_2), and reduced species such as methane (CH_4) and ammonia (NH_3) in the paleoatmosphere as O_2 evolved from 10^{-4} PAL to its present level. Our information concerning the chemical composition of the paleoatmosphere during the evolution of O_2 is based on two recent studies: a detailed review of the available geological and paleontological evidence by Walker (1977a) and a computer simulation of the chemical evolution of the atmosphere by Hart (1978). Walker concluded that the atmosphere formed via volatile outgassing very early in the Earth's history and that the paleoatmosphere contained about as much N_2, H_2O, and CO_2 as the present atmosphere. Hart's computer simulation developed the following scenario: The O_2 released from the photodissociation of H_2O and from photosynthesis (after the first 800 million years) chemically destroyed the CH_4 and NH_3 in the paleoatmosphere. By roughly 2 billion years ago, all but the trace amounts of reduced gases has been removed from the atmosphere, and at that point the atmosphere consisted primarily of N_2 (about 96%). Hart's calculations indicate that both CH_4 and NH_3 reached their present atmospheric levels about 2 billion years ago. The studies of Walker and Hart suggest that during the evolution of O_2 and O_3, the chemical composition of the paleoatmosphere was similar to the composition of the present atmosphere. Due to the similarity (with the exception of O_2 and O_3) in the composition of the paleoatmosphere and the modern atmosphere, we have adopted the O_3 photochemical and chemical reactions used in the current investigations of possible inadvertent depletion of O_3 due to anthropogenic activities (Hudson, 1977). The chemistry of CH_4, NH_3, CO_2, CO, and the chlorine species is not included in our calculations. The photochemical and chemical reactions used in our model are listed in Tables I and II.

To calculate the photodissociation rates of the molecular species given in Table I, the solar spectrum between 110 and 735 nm was divided into 174 spectral intervals, with molecular cross sections for each species folded into these spectral intervals. The solar flux data are from Ackerman (1971). The species absorption cross section references are also given in Table I. The calculations of the transmittance and rate of dissociation of molecular oxygen in the Schumann–Runge band (19 spectral intervals between 175 and 205 nm) are based on the data of Hudson and Mahle (1972). The Hudson and Mahle data include the values of band oscillator strengths and rotational line widths for the Schumann–Runge band system from which the transmittance and rate of dissociation of molecular oxygen as functions of temperature and oxygen column density have been calculated. For all of the photodissociation rates, the incident solar flux is attenuated by O_2, O_3, H_2O, CO_2, and CH_4 absorption and is calculated in 1-km altitude intervals

TABLE I
Photochemical reactions

No.	Photochemical reaction (s^{-1})	Reference for cross sections
J1	$O_2 + h\nu$ (110 – 175 nm) $\to O + O(^1D)$	Ackerman (1971), Watanabe (1958)
J2	$O_2 + h\nu$ (175 – 205 nm) $\to O + O$	Hudson and Mahle (1972a, b)
J3	$O_2 + h\nu$ (205 – 242 nm) $\to O + O$	Ackerman (1971), Hasson and Nicholls (1971)
J4	$O_3 + h\nu$ (110 – 310 nm) $\to O_2(^1\Delta_g) + O(^1D)$	Ackerman (1971), Inn and Tanaka (1953)
J5	$O_3 + h\nu$ (310 – 360 nm) $\to O_2(^1\Delta_g) + O$	Ackerman (1971), Inn and Tanaka (1953), Griggs (1968)
J6	$O_3 + h\nu$ (360 – 735 nm) $\to O_2 + O$	Ackerman (1971), Inn and Tanaka (1953)
J7	$H_2O + h\nu$ (110 – 200 nm) $\to OH + H$	Watanabe and Zelikoff (1953)
J8	$N_2O + h\nu$ (110 – 315 nm) $\to N_2 + O(^1D)$	Bates and Hays (1967), Johnston and Selwyn (1975)
J9	$HNO_3 + h\nu$ (110 – 240 nm) $\to H + NO_3$	Johnston and Graham (1974), Schmidt et al. (1974)
J10	$HNO_3 + h\nu$ (240 – 325 nm) $\to OH + NO_2$	Johnston and Graham (1974), Schmidt et al. (1974)
J11	$NO_2 + h\nu$ (110 – 245 nm) $\to NO + O(^1D)$	Dixon (1940), Hall and Blacet (1952), Nakayama et al. (1959)
J12	$NO_2 + h\nu$ (245 – 398 nm) $\to NO + O$	Dixon (1940), Hall and Blacet (1952), Nakayama et al. (1959)
J13	$H_2O_2 + h\nu$ (110 – 370 nm) $\to OH + OH$	Schürgers and Welge (1968), Paukert and Johnston (1972)

between the surface and 80 km. All of the photodissociation calculations are diurnal averages for a specified latitude and solar declination based on the procedure of Rundel (1977).

In the model, the following species profiles are calculated using a time-independent or steady-state species continuity equation, which combines the effects of both chemistry and vertical eddy transport: O_3, nitrous oxide (N_2O), and the odd nitrogen species (NO_x), which we define as the sum of nitric oxide (NO), nitrogen dioxide (NO_2), and nitric acid (HNO_3). The vertical distribution of the rapidly reacting atmospheric species (O, O(^{1}D), H, OH, HO_2, and H_2O_2) is determined solely by chemistry, which for these species is considerably faster than transport. The vertical distribution of the following species are specified as input parameters: H_2O (London and Park, 1974), CO_2 (Stewart and Hoffert, 1975), and CH_4 (Wofsy, 1976; Liu and Donahue, 1974).

In the present atmosphere the H_2O vapor mixing ratio above the tropopause is controlled by the tropopause temperature — the so-called 'cold trap'. A tenfold increase in the H_2O vapor mixing ratio above the tropopause requires a tropopause temperature increase of 15 K (Visconti, 1977). It does not appear that the tropopause temperature is strongly affected by even large variations in O_3 (Manabe and Strickler, 1964). To determine the

TABLE II
Chemical reactions

No.	Reaction	Rate constant ($cm^3 \, s^{-1}$ or $cm^6 \, s^{-1}$)	Reference
1	$O + O_2 + M \rightarrow O_3 + M$	$1.1 \times 10^{-34} \exp(510/T)$	Huie et al. (1972)
2	$O + O_3 \rightarrow 2O_2$	$1.9 \times 10^{-11} \exp(-2300/T)$	CIAP Monograph I (1974)
3	$O(^1D) + O_3 \rightarrow 2O_2$	1.2×10^{-10}	Hudson (1977)
4	$O(^1D) + M \rightarrow O + M$	$2.0 \times 10^{-11} \exp(107/T)$	Hudson (1977)
5	$N_2O + O(^1D) \rightarrow 2NO$	5.5×10^{-11}	Hudson (1977)
6	$N_2O + O(^1D) \rightarrow N_2 + O_2$	5.5×10^{-11}	Hudson (1977)
7	$NO + O + M \rightarrow NO_2 + M$	$1.55 \times 10^{-32} \exp(584/T)$	Hudson (1977)
8	$NO + O_3 \rightarrow NO_2 + O_2$	$2.1 \times 10^{-12} \exp(-1450/T)$	Hudson (1977)
9	$NO_2 + O \rightarrow O_2 + NO$	9.1×10^{-12}	Hudson (1977)
10	$NO_2 + O_3 \rightarrow NO_3 + O_2$	$1.2 \times 10^{-13} \exp(-2450/T)$	Hudson (1977)
11	$NO + HO_2 \rightarrow NO_2 + OH$	8×10^{-12}	Hudson (1977)
12	$NO_2 + OH + M \rightarrow HNO_3 + M$	$2.76 \times 10^{-13} \exp(880/T)/$ $(1.17 \times 10^{18} \exp(222/T + [M]))$	Hudson (1977)
13	$HNO_3 + OH \rightarrow NO_3 + H_2O$	8×10^{-14}	Hudson (1977)
14	$H_2O + O(^1D) \rightarrow 2OH$	2.3×10^{-10}	Hudson (1977)
15	$H + O_2 + M \rightarrow HO_2 + M$	$2.1 \times 10^{-32} \exp(290/T)$	Hudson (1977)
16	$H + O_3 \rightarrow OH + O_2$	$1.2 \times 10^{-10} \exp(-560/T)$	Hudson (1977)
17	$OH + O \rightarrow H + O_2$	4.2×10^{-11}	Hudson (1977)
18	$OH + O_3 \rightarrow HO_2 + O_2$	$1.5 \times 10^{-12} \exp(-1000/T)$	Hudson (1977)
19	$OH + OH \rightarrow H_2O + O$	$1 \times 10^{-11} \exp(-550/T)$	Hudson (1977)
20	$HO_2 + O \rightarrow OH + O_2$	3.5×10^{-11}	Hudson (1977)
21	$HO_2 + O_3 \rightarrow OH + 2O_2$	$7.3 \times 10^{-14} \exp(-1275/T)$	Hudson (1977)
22	$HO_2 + OH \rightarrow H_2O + O_2$	3×10^{-11}	Hudson (1977)
23	$HO_2 + HO_2 \rightarrow H_2O_2 + O_2$	2.5×10^{-12}	Hudson (1977)
24	$H_2O_2 + OH \rightarrow HO_2 + H_2O$	$1 \times 10^{-11} \exp(-750/T)$	Hudson (1977)
25	$H_2O_2 + O \rightarrow OH + HO_2$	$2.75 \times 10^{-12} \exp(-2125/T)$	Hudson (1977)

sensitivity of evolving O_3 to the choice of a H_2O vapor profile, we have performed calculations for the present atmospheric H_2O vapor mixing ratio and for H_2O vapor profiles equal to one-tenth and 10 times the present H_2O vapor profile of London and Park (1974). Since it is beyond the scope of this study to evaluate the eddy diffusion coefficient profile of the paleoatmosphere (there is some debate in the literature concerning the eddy diffusion profile in the present atmosphere), we have used the profile of McElroy et al. (1974) in all of the calculations presented here.

Another specified input parameter is the temperature profile of the paleoatmosphere. Walker (1977a) concluded that in the absence of O_3, the paleoatmosphere had a troposphere much like the present one and a more or less isothermal stratosphere and mesosphere. Initially, following the procedure of Ratner and Walker (1972), we did not attempt to evaluate temperature profiles in the stratosphere and mesosphere as O_2 and O_3 built up to present levels. Instead, we used two limiting cases. We used the temperature profile of the U.S. Standard Atmosphere (mid-latitude spring/fall temperature profile) for oxygen levels equal to or greater than 10^{-1} of the present atmospheric level (PAL)

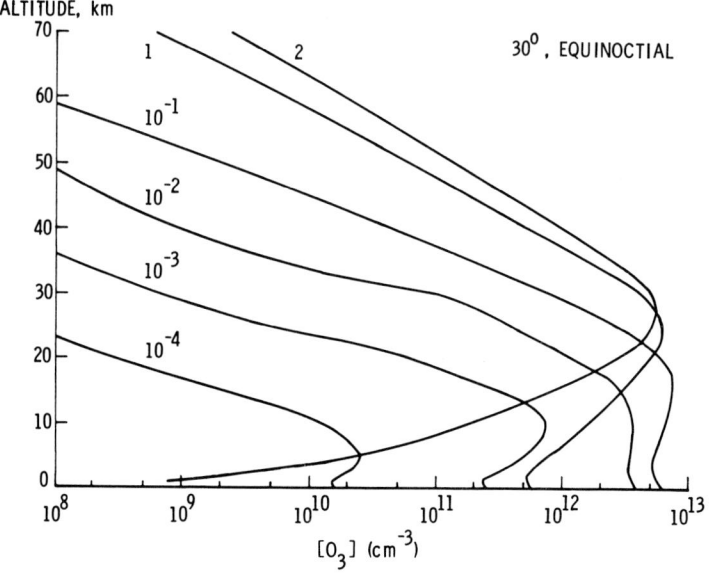

Fig. 1. The vertical distribution of O_3 as a function of atmospheric O_2 level.

of O_2 and a 'primordial' temperature profile for O_2 levels less than 10^{-1} PAL. The 'primordial' temperature linearly decreases from the tropopause (15 km) to the mesopause (90 km), resulting in an almost isothermal stratosphere. As noted by Ratner and Walker (1972), the effectiveness of O_3 absorption in producing a stratospheric temperature increase becomes smaller as the O_3 layer moves down to higher pressure levels and is negligibly small for O_3 density profiles corresponding to $O_2 \leqslant 10^{-1}$ PAL. The actual effect of paleoatmospheric O_3 on surface and atmospheric temperature was investigated by coupling the photochemical model with a radiative-convective model (Levine and Boughner, 1979). The results of this investigation are summarized in Section 5.

3. The Evolution of O_3

The vertical distribution of O_3 as O_2 evolved from 10^{-4} PAL to 1 PAL and for an O_2 level of 2 PAL for 30° latitude and equinoctial conditions is shown in Figure 1. Berkner and Marshall (1965) speculated that atmospheric O_2 levels may have exceeded 1 PAL before the present atmospheric level was achieved. We see that as the O_2 level increased from 10^{-4} PAL to 1 PAL, the height of the O_3 peak moved from about 5 km to about 25 km. Our calculations indicate that maximum O_3 densities at the surface and through the troposphere were achieved for an O_2 level of 10^{-1} PAL. We calculate surface and tropospheric O_3 densities of about 5×10^{12} cm^{-3} for an O_2 level of 10^{-1} PAL compared to densities of about 5×10^{11} cm^{-3} in the present atmosphere.

The large surface and tropospheric O_3 densities (about 5×10^{12} cm^{-3}) in the paleo-atmosphere found in this study were not found in the previous studies. Ratner and Walker

(1972) reported maximum surface O_3 densities of about 1×10^{12} cm^{-3} based on their calculations using the Chapman reactions. Hesstvedt et al. (1974) found maximum surface O_3 densities of less than 3×10^{10} cm^{-3} for calculations assuming photochemical equilibrium (no transport) and found maximum surface O_3 of about 7×10^{11} cm^{-3} with the identical chemistry, but including vertical eddy transport. Blake and Carver (1977), assuming photochemical equilibrium in their study, reported maximum surface O_3 densities of 2×10^{12} cm^{-3} for the present atmosphere and smaller surface O_3 levels for reduced O_2 levels. The enhanced surface and tropospheric O_3 densities calculated in our study are due in part to the inclusion of vertical eddy diffusion. The importance of vertical eddy transport on the distribution of O_3, particularly below the O_3 peak, has been discussed by Nicolet (1975). The importance of vertical eddy transport is also clearly seen in the calculations of Hesstvedt et al. (1974). Using identical chemical schemes, they reported an increase in surface O_3 of more than a factor of 20 for an O_2 level of 10^{-1} PAL when vertical eddy transport was included in their calculations compared to their photochemical equilibrium calculations (no transport).

Of all tropospheric species, O_3 comes closest to being naturally present at toxic levels (Chameides and Walker, 1975). Many varieties of plant life are extensively damaged when exposed to O_3 concentrations only two or three times greater than the present average ambient concentrations. Chameides and Walker (1975) examined the possible variation of tropospheric O_3 over geological time. They considered how changes in the CH_4 production rate over geological time would affect the production of O_3, and concluded that a tenfold increase in the CH_4 production rate would cause a fourfold increase in the tropospheric concentration of O_3 for the present level of O_2. Our calculations indicate much larger tropospheric levels of O_3 than calculated by Chameides and Walker corresponding to an O_2 level of 10^{-1} PAL. The toxic effects of these enhanced levels of tropospheric O_3 may have had significant adverse effects on both animal and plant life.

The evolution of the total O_3 column above the Earth's surface as a function of O_2 level is shown in Figure 2. The calculation of Berkner and Marshall (1965) is shown as the broken line curve 1, and our calculation for 30° latitude and equinoctial conditions is shown as the solid line curve 2. The maximum in total O_3 column for an O_2 level of 10^{-1} PAL, which is contrary to the Berkner and Marshall results, was first pointed out by Ratner and Walker, and later confirmed by Blake and Carver. The O_3 maximum for an O_2 level of 10^{-1} PAL resulted from the deeper penetration of solar ultraviolet radiation responsible for the production of O via the photodissociation of O_2. The enhanced number of third bodies (M) at the lower altitude increased the rate of formation of O_3 via the three-body reaction: $O + O_2 + M \rightarrow O_3 + M$, without a corresponding increase in the rate of destruction of O_3.

We have to re-emphasize that all of the calculations of O_3 in the O_2-deficient paleoatmosphere described in Levine et al. (1979) and discussed in this paper (Figures 1 and 2) were obtained with a stratospheric photochemical model. Several tropospheric chemical and photochemical processes which may affect the evolution of O_3 in the O_2-deficient paleoatmosphere not included in Levine et al. (1979) are the methane oxidation chain,

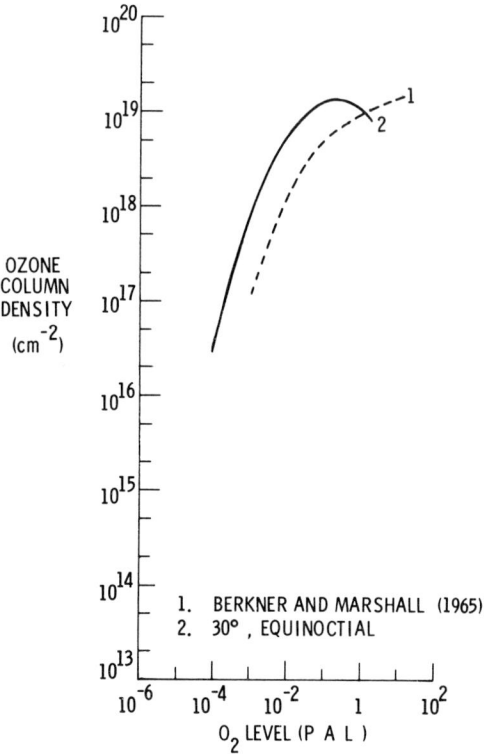

Fig. 2. The total O_3 column above the Earth's surface as a function of atmospheric O_2 level: comparison of results of Berkner and Marshall (1965) with the present study. Curve 1 is from Berkner and Marshall; Curve 2 is for 30° latitude, equinoctial conditions.

the rainout of soluble tropospheric species, e.g., HNO_3 and H_2O_2, realistic surface fluxes of biogenic trace species, e.g., N_2O and CH_4 and the inclusion of the surface as a physical sink for O_3. In addition, the stratospheric model of Levine et al. (1979) did not include the chemistry of the carbon and chlorine species.

4. The Variation of Solar Ultraviolet Radiation Incident at the Earth's Surface

Life on Earth has evolved under a shield of O_3 which absorbs biologically harmful solar ultraviolet (UV) radiation in the wavelength range from 200 to 300 nm. The critical stages in biological evolution are likely to have been linked to the thickness of the ozone shield (Berkner and Marshall, 1965; Ratner and Walker, 1972). We have calculated the variation of UV radiation incident at the Earth's surface in this spectral interval as O_3 evolved to its present atmospheric level by considering the attenuation of the solar UV due to the absorption by O_3, O_2, H_2O, CO_2, and CH_4 (Levine, 1978). For these calculations we divided the solar spectrum into 34 spectral intervals between 200 and 300 nm using the solar flux data of Ackerman (1971). The absorption cross sections for O_3,

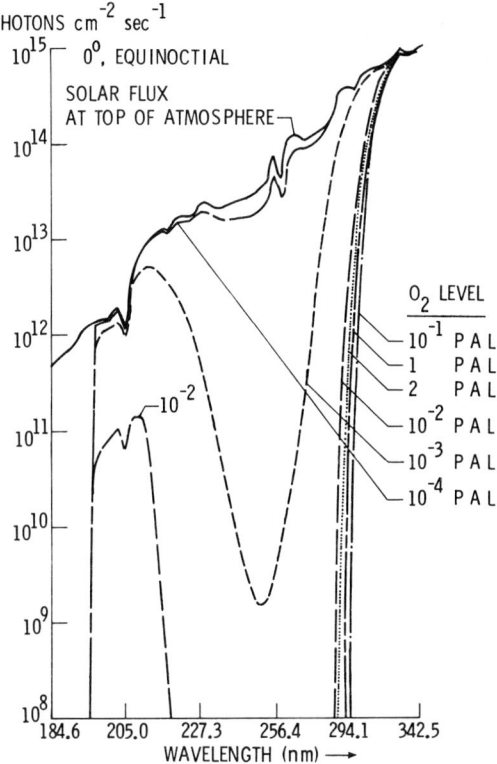

Fig. 3. The solar ultraviolet radiation between 184.6 and 342.5 nm reaching the Earth's surface for O_3 levels corresponding to O_2 levels ranging from 10^{-4} to 2 PAL for a lattitude of 0° and a solar declination of 0° (equinoctial conditions).

O_2, H_2O, CO_2, and CH_4 were folded into these spectral intervals. For each calculation, the O_2 level was specified and the calculated O_3 profile shown in Figure 1 was used. Present atmospheric levels of H_2O, CO_2, and CH_4 were used in all calculations. The results of these calculations for equinoctial conditions (the Sun on the Equator) for three different latitudes – 0°, 30°, and 57.3° – are shown in Figures 3 to 5. In these figures we have plotted the solar UV flux at the top of the atmosphere and at the surface in photons cm^{-2} s^{-1} for O_2 levels ranging from 10^{-4} to 2 PAL. Examination of these figures shows the existence of an UV window between 200 and 220 nm that closed only when O_2 reached 10^{-2} PAL (Levine, 1978). The effect of latitude (solar zenith angle) in determining the UV radiation incident at the surface is clearly seen in these figures.

5. Paleoatmospheric O_3 and Surface and Atmospheric Temperature

In this section we assess the influence of O_3 in determining the surface temperature of the paleoatmosphere, prior to the build-up of molecular oxygen (O_2) to its present

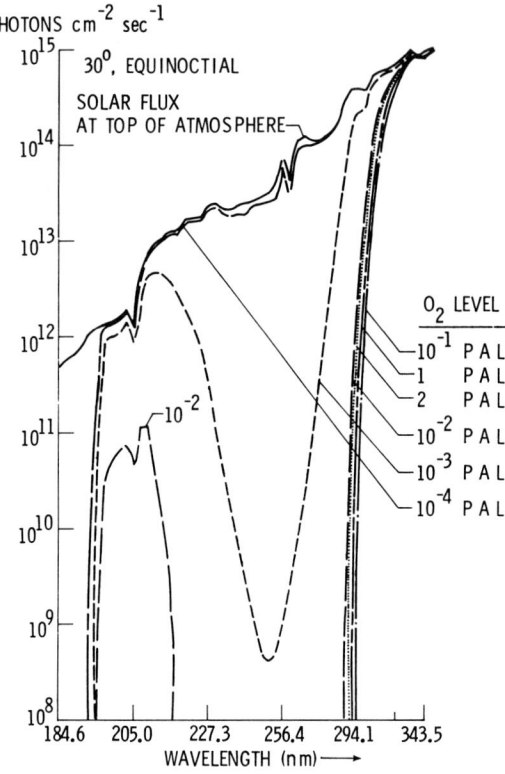

Fig. 4. The solar ultraviolet radiation between 184.6 and 342.5 nm reaching the Earth's surface for O_3 levels corresponding to O_2 levels ranging from 10^{-4} to 2 PAL for a latitude of 30° and a solar declination of 0° (equinoctial conditions).

atmospheric level (PAL). Previous studies have either dismissed the radiative effects of O_3 in determining the surface temperature of the paleoatmosphere and have invoked the presence of very large amounts of ammonia (about 10 ppm of NH_3) (Sagan and Mullen, 1972; Sagan, 1977) or have invoked the presence of large amounts of carbon dioxide ranging from factors of 3.5 to 10 (Morss and Kuhn, 1978) to several orders of magnitude more CO_2 than in the present atmosphere (Owen et al., 1979).

The surface temperature and tropospheric and stratospheric temperature profile of the paleoatmosphere were calculated using a radiative-convective model similar to that described by Ramanathan (1976). This model employs a combination of band-absorptance and emissivity formulations to account for radiative effects of CO_2, H_2O, and O_3. Expressions for the band absorptance and emissivity and the specific infrared and solar absorption bands of O_3, H_2O, and CO_2 used in our calculations were taken from Ramanathan (1976). The effects of clouds were taken into account by using the procedure described by Ramanathan et al. (1976); that is, by using an equivalent, single-layer cloud model. The fractional cloud amount and the cloud reflectivity to solar radiation were taken to be 0.446 and 0.515, respectively. The ground reflectivity was assumed to be

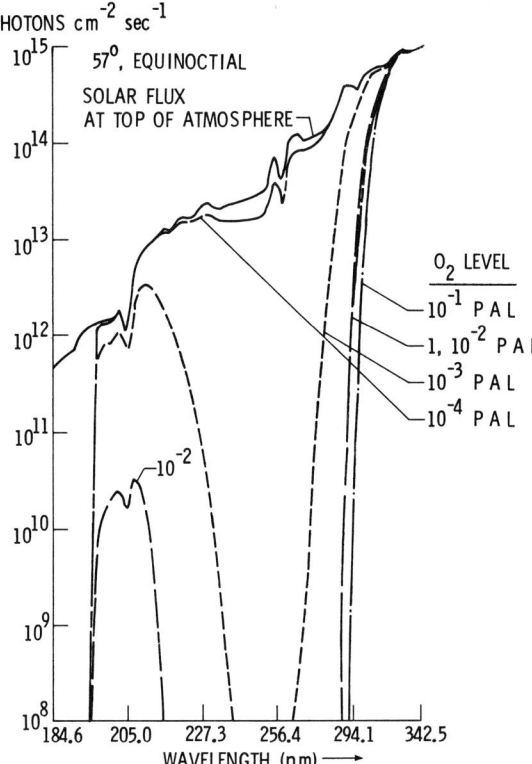

Fig. 5. The solar ultraviolet radiation between 184.6 and 342.5 nm reaching the Earth's surface for O_3 levels corresponding to O_2 levels ranging from 10^{-4} to 1 PAL for a latitude of 57° and a solar declination of 0° (equinoctial conditions).

0.105. In the present study, the cloud height was taken to be 5 km, which was held fixed in all our calculations. A discussion concerning the sensitivity of the computed surface temperature to the assumed cloud model is in Ramanathan et al. (1976). The tropospheric temperature lapse rate was assumed equal to -6.5 K km^{-1}, a mid-latitude continental lapse rate. The computational procedures used to obtain the surface temperature and atmospheric temperature profile were similar to those outlined by Ramanathan (1976). Changes in planetary albedo due to variations in ozone content were also incorporated in our calculations in an approximate manner by parameterizing the results of Ramanathan et al. (1976), which are based on detailed multiple scattering calculations. Our calculations also consider the albedo changes associated with variations in water vapor amount.

To assess the maximum influence of paleoatmospheric O_3 on surface temperature, we used the O_3 level corresponding to an O_2 level of 10^{-1} PAL, the maximum total O_3 column over geological history, according to the calculations of Levine et al. (1979). In addition, because ozone production and loss rates are sensitive to temperature, feedback between the temperature field and ozone concentration was taken into account by iterating between the photochemical and radiation codes so that the computed ozone and temperature profiles were consistent with each other.

The vertical temperature profile due to the O_3 profile for an O_2 level of 10^{-1} PAL is compared with the calculated temperature for present-day conditions (O_2 = 1 PAL) in Figure 6. Both calculations were performed for the present solar luminosity, as were all of the calculations of Morss and Kuhn (1978). Due to the decreased stratospheric O_3 concentrations for the 10^{-1} PAL case and a corresponding reduction in the absorption of solar energy, stratospheric temperatures are substantially lower in comparison to present atmospheric conditions at altitudes greater than about 22 km, the location of the maximum O_3 mixing ratio for the 10^{-1} PAL case. Below 22 km, O_3 mixing ratios are significantly larger, which, in conjunction with the reduced O_3 column density above, results in enhanced solar heating (approximately a factor of 4 larger at 16 km), and hence higher temperatures compared with the present atmosphere. The flattening of the temperature profile between 15 and 30 km represents the net balance between the absorption of solar and surface radiation by O_3 and infrared cooling by H_2O and CO_2.

The radiative effects of O_3 — both its altitude redistribution and the enhanced total — corresponding to an O_2 level of 10^{-1} PAL, resulted in a globally-averaged surface temperature increase of about 4.5 K. This calculation was for surface pressure of 1 atmosphere, as were the calculations of Owen *et al.* (1979), i.e., the effect of reduced atmospheric mass on surface temperature was not taken into account. By maintaining a surface pressure of 1 atmosphere, our temperature calculations tend to emphasize the relative contribution of O_3 — due to its total column content and altitude distribution — to

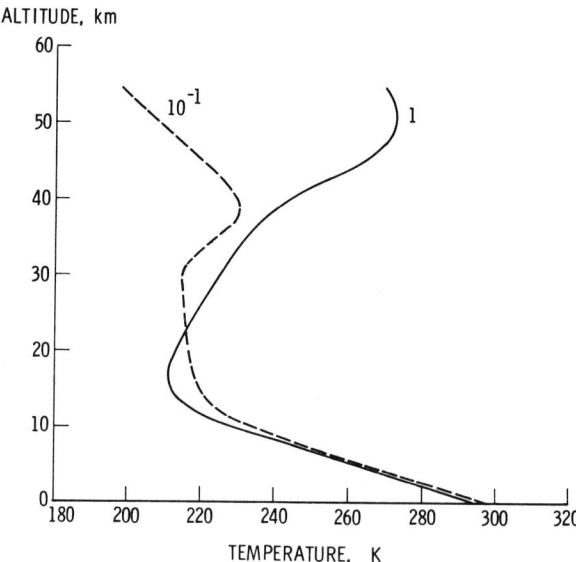

Fig. 6. The vertical distribution of temperature in the troposphere and stratosphere based on radiative–convective calculations for O_3 levels corresponding to O_2 levels of 1 and 10^{-1} PAL.

the computed surface temperature. As described later, calculations were also made for a reduced surface pressure, corresponding to an O_2 level of 10^{-1} PAL, which also alters radiative transport in the H_2O and CO_2 infrared bands. The increase in globally-averaged surface temperature that we calculate is the result of a combination of the following factors: an increased infrared greenhouse effect due to O_3, a decrease in visible albedo due to the enhanced O_3 column density, and greenhouse and albedo variations associated with changes in the H_2O concentration. Each of these factors will be briefly discussed.

The greenhouse effect of O_3 is basically a measure of the net absorption of surface radiation by the O_3 9.6 μm band. This absorption depends on both the total amount of O_3 and the 'effective broadening' pressure, whose importance in determining the outgoing flux within the 9.6 μm band was first discussed by Ramanathan *et al.* (1976). An increase in either one of these factors increases the absorption of surface radiation and, in order to maintain a radiative balance for the Earth-atmosphere system, requires the surface temperature to increase because the absorbed surface energy is re-emitted at a lower temperature. The effective broadening pressure is an O_3 mass-weighted average of pressure through the atmosphere, and gives greater weight to those atmospheric regions near the O_3 maximum. An increase in effective broadening pressure implies more absorption at lower altitudes where the pressure is higher and the absorption lines are broader. For the 10^{-1} PAL case shown in Figure 1, the O_3 column amount is 40% larger than the present-day case. In addition, as illustrated in Figure 2, there is a substantial lowering of the O_3 maximum so that there is approximately a factor of 3 increase in the effective broadening pressure. Our calculations indicate that about one-third of the surface temperature increase can be attributed to the enhanced greenhouse effect of O_3; of this fraction, approximately half is due to the increase in effective broadening pressure, i.e., enhanced absorption at lower altitudes.

As previously mentioned, an increase in O_3 increases the absorption of radiation from the direct solar beam and decreases the amount of solar radiation reflected from the lower atmosphere and surface escaping to space, i.e., decreases the visible albedo. Both processes lead to greater deposition of solar energy in the Earth-atmosphere system and to a warming. Calculations indicate that the increase in O_3 and H_2O vapor absorption in the visible is responsible for approximately two-thirds of the surface temperature increase.

An additional calculation was performed to study the radiative effects of O_3 for a reduced surface pressure of 0.81 atmosphere, corresponding to the surface pressure for an O_2 level of 10^{-1} PAL. We found the surface temperature approximately 5 K lower compared to the calculation performed for 1 atmosphere. The lower temperature for the 0.81 surface pressure resulted from the combined effects of reduced line broadening on the H_2O, and CO_2 infrared bands, which affects the greenhouse enhancement, and the reduced H_2O and CO_2 amounts which affects both the planetary albedo and the infrared greenhouse enhancement. We found that the reduced line broadening of the H_2O and CO_2 infrared bands and the reduced H_2O and CO_2 amounts for the 0.81 surface pressure case appear to compensate for the temperature increase due to the amount and distribution of O_3 for the 10^{-1} PAL of O_2. However, the relative effect of O_3 on surface temperature remains about the same. In other words, the O_3 corresponding to an O_2 level of 10^{-1} PAL increases the surface temperature by approximately 4.5 K above what it would be if the O_3 distribution were the same as it is in the present atmosphere (1 PAL of O_2).

In conclusion, we have found that the radiative effects, at infrared and solar wavelengths, of ozone for an O_2 level of 10^{-1} PAL resulted in a globally-averaged surface temperature increase of about 4.5 K for the present solar constant. The precise implications of these results on the paleoclimate are not clear at the present time, due to considerable uncertainty concerning the chronology for the evolution of O_2. According to differing evolution scenarios, O_2 may have reached 10^{-1} PAL as early as 2 billion years ago (Walker, 1977b), or as late as 400 million years ago (Berkner and Marshall, 1965). It should be noted that a 4.5 K mean surface temperature increase is approximately 2.2 times larger than the most recently calculated temperature rise due to a doubling of the carbon dioxide concentration in the present atmosphere (Augustsson and Ramanathan, 1977).

The evolution of atmospheric O_3 over geological time controlled the solar ultraviolet radiation reaching the Earth's surface and affected the temperature of the surface and atmosphere. All of the calculations of O_3 in the O_2-deficient paleoatmosphere presented here were obtained with a stratospheric photochemical model (Levine *et al.*, 1979). Future studies of paleoatmospheric O_3 should include tropospheric photochemical and chemical processes, as well as the chemistry of the carbon and chlorine species. To-date, these processes and species have not been included in any study of O_3 in the O_2-deficient paleoatmosphere.

References

Ackermann, M.: 1971, in G. Fiocco (ed.), *Mesospheric Models and Related Experiments*, D. Reidel Publ. Co., Dordrecht, Holland, pp. 149–159.
Augustsson, T. and Ramanathan, V.: 1977, *Journal of the Atmospheric Sciences* **34**, 448.
Bates, D. R. and Hays, P. B.: 1967, *Planetary and Space Science* **15**, 189.
Berkner, L. V. and Marshall, L. C.: 1965, *Journal of the Atmospheric Sciences* **22**, 225.
Blake, A. J. and Carver, J. H.: 1977, *Journal of the Atmospheric Sciences* **34**, 720.
Chameides, W. and Walker, J. C. G.: 1975, *American Journal of Science* **275**, 737.
CIAP Monograph I: 1974, *The Natural Stratosphere of 1974*. Final Report prepared for the Climatic Impact Assessment Program, Dept. of Transportation, Washington, D.C.
Crutzen, P. J.: 1970, *Quarterly Journal of the Royal Meteorological Society* **96**, 320.
Dixon, J. K.: 1940, *Journal of Chemical Physics* **8**, 157.
Griggs, M.: 1968, *Journal of Chemical Physics* **49**, 857.
Hall, T. C. and Blacet, F. E.: 1952, *Journal of Chemical Physics* **20**, 1745.
Hart, M. H.: 1978, *Icarus* **33**, 23.
Hasson, V. and Nicholls, R. W.: 1971, *Proceedings Physical Society London Atomic and Molecular Physics* **4**, 1789.
Hesstvedt, E., Henriksen, S. E., and Hjartarson, H.: 1974. *Geophysica Norvegica* **31**, 1.
Hudson, R. D.: 1977, *Chlorofluoromethanes and the Stratosphere*. National Aeronautics and Space Administration Reference Publication 1010.
Hudson, R. D. and Mahle, S. H.: 1972, *Journal of Geophysical Research* **77**, 2902.
Huie, R. E., Herron, J. T., and Davis, D. D.: 1972, *Journal of Physical Chemistry* **76**, 2653.
Inn, E. C. Y. and Tanaka, Y.: 1953, *Journal of the Optical Society of America* **43**, 870.
Johnston, H. S. and Graham, R.: 1974, *Canadian Journal of Chemistry* **52**, 1415.

Johnston, H. S. and Selwyn, G. S.: 1975, *Geophysical Research Letters* **2**, 549.
Levine, J. S.: 1978, *EOS, Transactions American Geophysical Union* **59**, 341.
Levine, J. S. and Schryer, D. R. (Editors): 1978, *Man's Impact on the Troposphere*. National Aeronautics and Space Administration Reference Publication No. 1022.
Levine, J. S. and Boughner, R. E.: 1979, *Icarus* **39**, 310.
Levine, J. S., Hays, P. B., and Walker, J. C. G.: 1979, *Icarus* **39**, 295.
Liu, S. C. and Donahue, T. M.: 1974, *Journal of the Atmospheric Sciences* **31**, 1118.
London, J. and Park, J. H.: 1974, *Canadian Journal of Chemistry* **52**, 1599.
Manabe, S. and Strickler, R. F.: 1964, *Journal of the Atmospheric Sciences* **21**, 361.
Manabe, S. and Wetherald, R. T.: 1967, *Journal of the Atmospheric Sciences* **24**, 241.
Margulis, L. and Lovelock, J. E.: 1974, *Icarus* **21**, 471.
Mastenbrook, H. J.: 1971, *Journal of the Atmospheric Sciences* **28**, 1495.
McElroy, M. B. and McConnell, J. C.: 1971, *Journal of the Atmospheric Sciences* **28**, 1095.
McElroy, M. B., Wofsy, S. C., Penner, J. E. and McConnell, J. C.: 1974, *Journal of the Atmospheric Sciences* **31**, 287.
Morss, D. A. and Kuhn, W. R.: 1978, *Icarus* **33**, 40.
Nakayama, T., Kitamura, M. Y., and Watanabe, K.: 1959, *Journal of Chemical Physics* **30**, 1180.
Nicolet, M.: 1975, *Reviews of Geophysics and Space Physics* **13**, 593.
Owen, T., Cess R. D., and Ramanathan, V.: 1979, *Nature* **277**, 640.
Paukert, T. T. and Johnston, H. S.: 1972, *Journal of Chemical Physics* **56**, 2824.
Ramanathan, V.: 1976, *Journal of the Atmospheric Sciences* **33**, 1330.
Ramanathan, V., Callis, L. B., and Boughner, R. E.: 1976, *Journal of the Atmospheric Sciences* **33**, 1092.
Ratner, M. I. and Walker, J. C. G.: 1972, *Journal of the Atmospheric Sciences* **29**, 803.
Rundel, D. R.: 1977, *Journal of the Atmospheric Sciences* **34**, 639.
Sagan, C.: 1977, *Nature* **269**, 224.
Sagan, C. and Mullen, G.: 1972, *Science* **177**, 52.
Schmidt, S. G., Amme, R. C., Murcray, D. G., Goldman, A., and Bonomo, F. S.: 1974, *Nature* **238**, 109.
Schürgers, M. and Welge, K. H.: 1968, *Z. Naturforseh.* **A23**, 1508.
Stewart, R. W. and Hoffert, M. I.: 1975, *Journal of the Atmospheric Sciences* **32**, 195.
U.S. Standard Atmosphere: 1962, U.S. Government Printing Office, Washington, D.C.
Visconti, G.: 1977, *Journal of the Atmospheric Sciences* **34**, 193.
Walker, J. C. G.: 1977a, *Evolution of the Atmosphere*, MacMillan Publishing Co., New York.
Walker, J. C. G.: 1977b, *EOS, Transactions of the American Geophysical Union* **58**, 689.
Watanabe, K.: 1958, *Advances in Geophysics* **5**, 153.
Watanabe, K. and Zelikoff, M.: 1953, *Journal of the Optical Society of America* **43**, 753.
Wofsy, S. C.: 1976, *Annual Review, Earth and Planetary Sciences* **4**, 441.

ATMOSPHERIC CONSTRAINTS ON THE EVOLUTION OF METABOLISM

JAMES C. G. WALKER

Arecibo Observatory, National Astronomy and Ionosphere Center, Arecibo, Puerto Rico 00612

Abstract. Earth's early history may have been characterized by coevolution of microbial metabolism and atmospheric composition. Metabolic developments affected the composition of the atmosphere and the resultant changes in the atmosphere stimulated the evolution of new metabolic capabilities.

The first organisms were presumably fermenting heterotrophs, exploiting organic molecules abiotically synthesized. These organisms multiplied, developing new biosynthetic capabilities to overcome deficiencies in the abiotic supply of particular compounds, until their growth was limited by the energy source provided by abiotic synthesis of fermentable organic compounds. Further growth required a new energy source, which may have been the chemical energy represented by the mixture of carbon dioxide and hydrogen in the primitive atmosphere. Chemotrophic organisms resembling methane bacteria may have evolved to exploit this source. They would have flourished, along with the heterotrophs that fed on them, until they had decreased the level of atmospheric hydrogen to the point where further extraction of chemical energy from the atmosphere was not possible. Once again, the expansion of life was limited by the availability of energy.

The origin of bacterial photosynthesis overcame the second energy crisis. Photosynthetic bacteria could exploit the abundant energy of sunlight while using atmospheric hydrogen and reduced compounds derived from it only as electron donors. Life flourished again, drawing atmospheric hydrogen (replenished only by volcanoes) down to levels so low as to limit even bacterial photosynthesis. Before the full potential of photosynthesis could be exploited the evolution of the metabolic apparatus to process an electron donor of unlimited abundance was necessary. This donor, of course, was water, and the new metabolic process was algal photosynthesis. The oxygen released changed the world from anaerobic to aerobic and made possible the last great advance in energy-yielding metabolism, aerobic respiration.

1. Introduction

What does metabolism have to do with the atmosphere? Metabolism refers to the chemical processes within an organism that provide new cell material (biosynthesis) as well as the processes that provide the organism with the energy it needs for biosynthesis and other functions. The raw materials for this chemical activity are extracted from the surroundings, and waste products are dumped into the surroundings, so the metabolic activity of organisms changes the composition of the environment, including the atmosphere. The most outstanding example of this effect is the oxygen-rich atmosphere of earth, sustained in a reducing universe by the metabolic activities of algae and green plants. Figure 1 illustrates the aspects of metabolism that should be kept in mind.

In this paper I discuss how early organisms may have affected the composition of the primitive atmosphere. I must, therefore, consider the metabolic capabilities of early

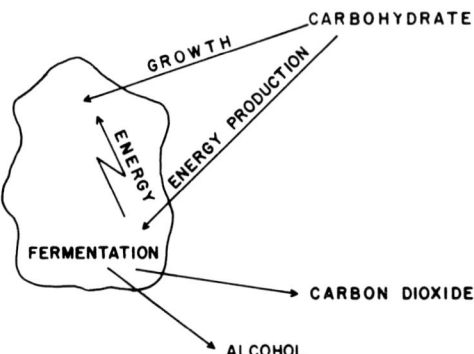

Fig. 1. Schematic representation of a microbe deriving energy from the fermentation of organic compounds in its environment and using the energy for growth by incorporating available compounds into its cellular material.

organisms and how these capabilities may have developed. There is no geological record of most of the developments I shall talk about, so my account will be speculative in the extreme. This is a subject that allows the imagination free rein. In an attempt to constrain my own imagination I have collected four guiding principles, presented in Figure 2. These principles are not original, of course, and as stated and applied here, they are grossly simplified, but they do impose a measure of order on an otherwise unconstrained problem.

Principle 1 states that biologically interesting gases move in cycles. If organisms are consuming some gas we can ask what source is maintaining the supply of that gas. If organisms are releasing some waste product to the environment we can ask what happens to that product. As an example drawn from the modern world we can take oxygen, the waste product produced by photosynthetic algae and green plants. It does not simply accumulate in the atmosphere. In fact, the rate of production is great enough to provide all the oxygen in the atmosphere in only 4000 yr. Atmospheric composition is stable because animals, including microbes, consume oxygen in the metabolic process of respiration almost as fast as it is produced. So, in thinking about the ancient world it is appropriate, even necessary, to consider sources and sinks of the biologically interesting gases and to devote attention to kinetic factors, the rates of production and destruction of these gases. As an aid to this kind of thinking I offer, in Figure 3, my view of the cycles of a number of important constituents of the prebiological atmosphere. The figure shows that I attach considerable importance to a volcanic source of atmospheric gases, particularly the reduced gases hydrogen and carbon monoxide which, because of the escape

1. BIOLOGICALLY INTERESTING GASES MOVE IN CYCLES
2. BIOLOGICAL CYCLES ARE LEAKY
3. SATISFIED CREATURES DO NOT CHANGE
4. ORGANISMS ARE GREEDY

Fig. 2. Guiding principles.

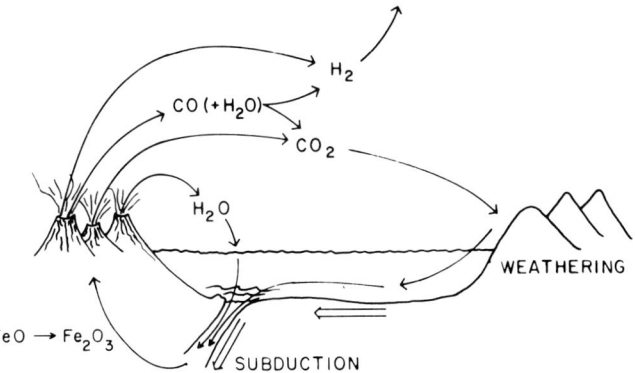

Fig. 3. Schematic representation of the cycles of important gases in the prebiological atmosphere.

of hydrogen to space, can survive in the atmosphere for only limited periods of time (Walker, 1977).

The example of oxygen in the modern atmosphere serves also to illustrate Principle 2. Although biological cycles are rapid in terms of geological spans of time, they are leaky. Even in the aerobic world of today not all of the organic carbon produced by photosynthetic organisms is consumed in respiration. Some of it survives to become buried in sediments on the floor of the ocean (see Figure 10). The global rate of respiration is therefore slightly less than the global rate of photosynthesis, and some oxygen leaks out of the biological cycle to be consumed by the weathering of reduced minerals in rocks (Van Valen, 1971; Holland, 1973a; Walker, 1974; Garrels *et al.*, 1976). Because biological cycles are leaky, they cannot, by themselves, sustain the habitability of the biosphere for geological periods of time. Geological processes of weathering, sea-floor spreading, tectonic activity, and volcanism are needed to recycle material that leaks out of the rapid biological cycles. In what follows I shall be keeping a close eye on the leaks.

2. Evolution of Biosynthetic Pathways

Life was preceded by the abiotic synthesis of organic compounds from the inorganic constituents of the atmosphere. For simplicity, we may think of abiotic synthesis as a process powered by lightning, sunlight, or meteor impact that converted atmospheric carbon dioxide (or carbon monoxide) and hydrogen into a variety of organic compounds. The first organisms exploited the products of abiotic synthesis. Of limited biosynthetic capability, they built themselves out of available organic compounds and derived energy by the metabolic process of fermentation, in which these same organic compounds served as electron donors and electron acceptors. No significant biological recycling of material was possible at this stage because fermentation produced waste products from which further energy could not be extracted by fermentation. These waste products presumably escaped into the atmosphere or settled to the bottom of the sea where they were removed from the biosphere (Figure 4).

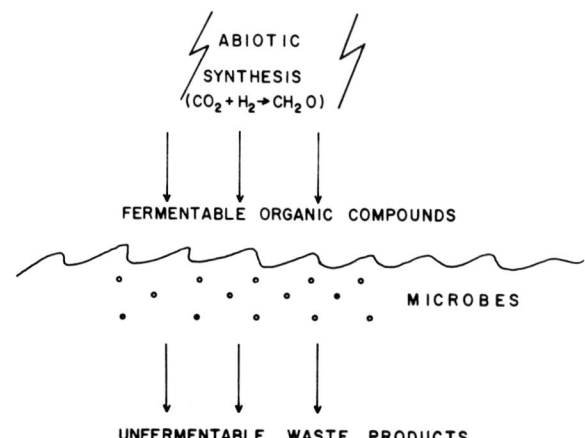
Fig. 4. The expansion of life was limited by the abiotic source of fermentable organic compounds.

In the light of Principle 3, "Satisfied creatures do not change", we may ask why life did not settle down at the stage illustrated by Figure 4. Principle 4 provides the answer: "Organisms are greedy". The first microbes ate and expanded in population until their further growth was limited by a shortage of some essential organic molecule. Abiotic synthesis prior to the origin of life may have provided a store of food for the very first organisms, but given the continual drain of organic waste products into sediments, there must have come a time when this store was exhausted, and further growth was limited by the rate of ongoing abiotic synthesis.

This was the time of elaboration of biosynthetic pathways in the primitive microbes. Population growth halted when some essential compound was being consumed by organisms as fast as it was being produced abiotically, until mutation produced an organism able to synthesize the needed compound out of other more abundant compounds. The new metabolic capability gave the mutant a competitive advantage, so it multiplied until the new capability was widespread. The total microbial population expanded until a different essential compound was in short supply. So mutation combined with natural selection caused biosynthetic pathways to evolve a step at a time in directions determined by the mixture of organic compounds furnished by abiotic synthesis (Miller and Orgel, 1974).

In time, however, the growth and development of life was brought up short by the first energy crisis. Energy for all of this biosynthetic activity was provided by fermentation of organic compounds abiotically synthesized. The overall level of biological activity was limited by the rate at which abiotic synthesis supplied fermentable organic compounds.

3. Energy Limited Ecosystems

Short of energy, our greedy creatures were no longer satisfied, so they changed. Mutation produced an organism able to tap the chemical energy represented by the mixture of inorganic oxidized and reduced gases in the atmosphere. Modern methane bacteria, which derive energy from the reaction of carbon dioxide and hydrogen to produce methane and water (cf. Stephenson, 1949; Stanier et al., 1970) exhibit the kind of metabolic capability I have in mind. For simplicity I assume that the ability to synthesize organic material from inorganic constituents (autotrophy) was an adjunct of the new metabolic process (chemoautotrophy). Freed at last from its dependence on abiotic synthesis, life flourished and population grew while the chemoautotrophs exploited abundant atmospheric stores of hydrogen and carbon dioxide and the heterotrophs exploited organic compounds synthesized by the autotrophs.

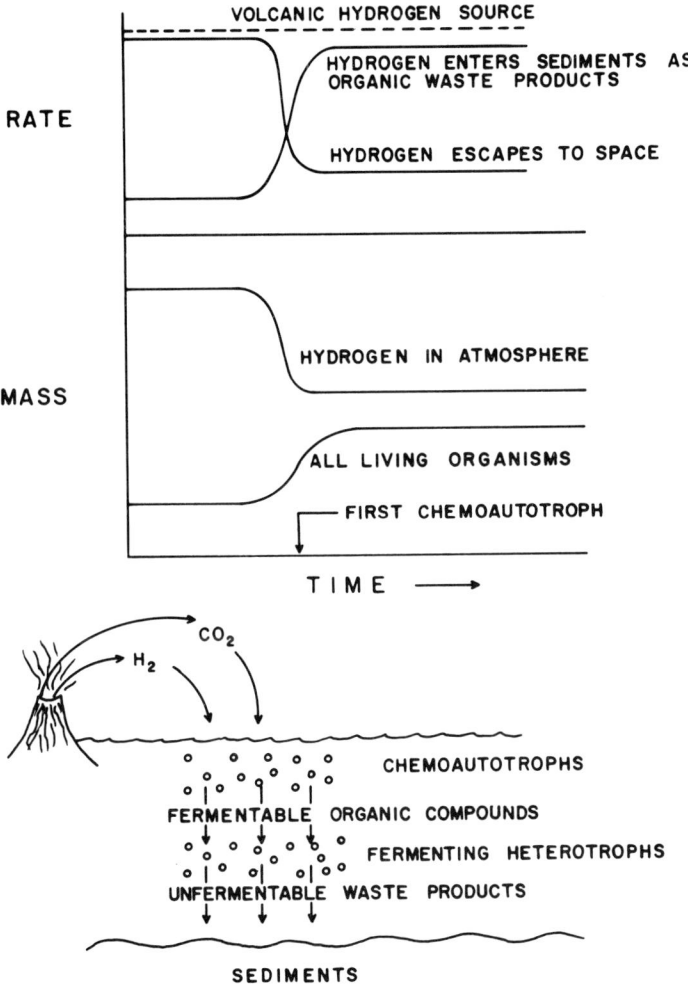

Fig. 5. Biomass increased after the introduction of chemoautotrophy, but the combined activities of chemoautotrophs and fermenting heterotrophs decreased the hydrogen content of the atmosphere, converting hydrogen that previously had escaped to space into organic waste products incorporated in sediments.

But the biological cycle was leaky, and the combined activities of autotrophs and heterotrophs extracted reduced and oxidized gases (which I call H_2 and CO_2 for convenience) from the atmosphere and converted them into unfermentable organic waste products that were incorporated into seafloor sediments (Figure 5). Life was greedy, of course, and expanded until something was in short supply. That something would have been either hydrogen or carbon dioxide, the raw materials that fueled the autotrophs. Since volcanoes release more carbon dioxide than hydrogen, I conclude that it was hydrogen that became the limiting factor (rates count more than abundances in accordance with Principle 1). The chemoautotrophs drew atmospheric hydrogen down to a level at which they could barely survive (it would be interesting to know what that level is for modern methane bacteria). Thereafter, biological productivity was limited by the rate of supply of reduced gas by volcanoes.

This event marked the beginning of a significant geochemical role for life. It was the first major change in atmospheric composition to be caused by organisms. That the biota was sufficiently active to change the atmosphere follows from Principle 3. If the environment had provided enough of what life wanted (hydrogen and carbon dioxide) there would have been no further biological change. The next development in metabolic capability eased the energy crisis caused by an inadequate supply of hydrogen.

4. Nutrient Limited Ecosystems

There evolved an organism possessing pigments, able to capture the energy of sunlight and use this energy for biosynthesis. Photosynthesis freed life from its dependence on volcanic hydrogen as a source of energy, but the first photosynthesizers, photosynthetic bacteria, still needed reduced compounds as electron donors for the reduction of carbon dioxide to organic molecules. A rapid biological cycle became possible for the first time, in which the unfermentable waste products of heterotrophs served as electron donors for bacterial photosynthesis, in the process being oxidized sufficiently to become useful to heterotrophs once again. A much expanded level of biological activity was possible, but the biological cycle was still leaky. Reduced compounds, derived ultimately from volcanic hydrogen, continued to drain out of the biosphere into sediments (Figure 6). Biological productivity was still limited by the rate of supply of volcanic hydrogen, but the hydrogen was used now not as an energy source but as a nutrient compound, an electron donor. Presumably photosynthetic bacteria can prosper at lower hydrogen partial pressures than chemoautotrophs, so the level of atmospheric hydrogen was reduced even further. Experimental evidence on the hydrogen requirements of photosynthetic bacteria as well as methane bacteria would be most interesting.

Some photosynthetic bacteria use hydrogen sulfide as electron donor, oxidizing it to the sulfate ion (Postgate, 1968). This capability, when it developed, may have provided the first abundant source of sulfate on earth, making possible the first new source of energy for heterotrophs, anaerobic respiration. Bacterial photosynthesis combined with anaerobic respiration provided a biological cycle for organic carbon much like the dominant

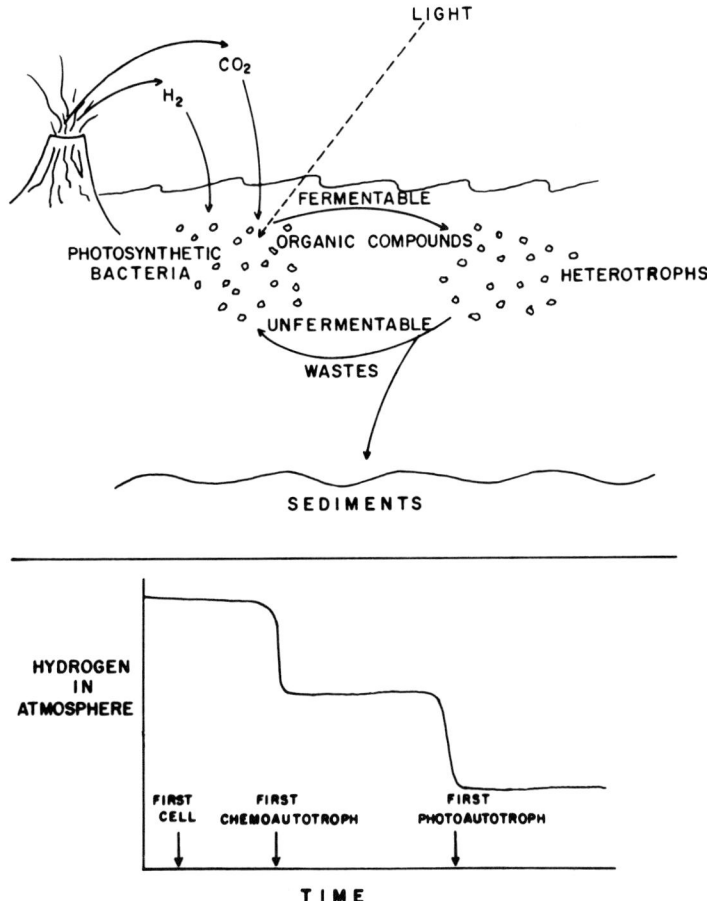

Fig. 6. Photoautotrophy caused another reduction in atmospheric hydrogen, but permitted the first rapid biological cycle of organic carbon.

modern cycle of algal photosynthesis and aerobic respiration (Figure 7). At this conference, however, Schidlowski (1979) presented isotopic evidence for the origin of sulfate reduction (anaerobic respiration) as recently as 2.8 billion years ago.

In spite of these improvements in the efficiencies of biological cycles, shortage of hydrogen continued to be a problem until life broke the hydrogen habit by learning to use abundant water as electron donor in photosynthesis. Algal photosynthesis was a considerable metabolic achievement, both because water is hard to oxidize (Olson, 1970) and because the waste product, oxygen, is potentially damaging to cell material (Fridovich, 1975, 1978). Once these difficulties had been overcome, however, the dependence of life on volcanic hydrogen ceased. Life was able to expand until some new factor became limiting. The new factor may have been phosphorus, the essential nutrient element that limits productivity over much of the modern ocean (Redfield, 1958; Redfield et al., 1963; Broecker, 1971, 1974).

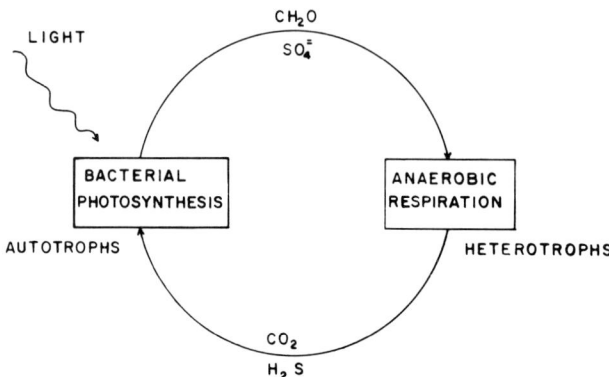

Fig. 7. Bacterial photosynthesis and anaerobic respiration sustained a biological cycle of organic carbon much like that of the modern day (see Figure 10), except that hydrogen sulfide and sulfate substituted for water and oxygen.

The presence of banded iron formation in the 3.8 billion year old rocks at Isua in West Greenland (Allaart, 1976), is evidence of the prior evolution of algal photosynthesis (Cloud, 1976). It appears that all of the developments I have so far described took place during the first 800 million years of Earth's history, a period from which no geological record has yet been discovered. Banded iron formations appear to have been deposited in an anaerobic world, but one possessed of a source of oxygen (MacGregor, 1927). The anaerobic world was required to permit transport of iron in solution in its ferrous form, but the oxygen source was required to precipitate the iron in its insoluble, ferric form (Cloud, 1973; Holland, 1973b; Walker, 1978a). The continued presence of banded iron formations in the geological record until about 2 billion years ago, along with other evidence (Cloud, 1968, 1972; Margulis et al., 1976), suggests that the world remained anaerobic for almost 2 billion years after the origin of algal photosynthesis.

Remembering the importance of kinetic considerations, I assume that for all of this time the rate of production of oxygen by algae remained less than the rate of supply of reduced species to the ocean as a result of the weathering of rocks combined with the rate of release of volcanic hydrogen (Figure 8). What limited the rate of algal photosynthesis to values much below those of the present day is an intriguing question. Whatever mechanism it was must not have been so effective as to prevent the eventual rise of the oxygen source above the source of reduced species or the world would still be anaerobic. One possibility (Walker, 1978b) is that the activities of algae were restricted originally by competition, presumably for phosphorus, by photosynthetic bacteria (Figure 9). The bacteria, with their energetically less expensive metabolic processes, may have preempted phosphorous as long as the supply of reduced electron donors was adequate. Recent evidence (Cohen et al., 1975; Garlick et al., 1977; Oren et al., 1977), in fact, reveals that many blue-green algae practice photosynthesis without release of oxygen if reduced electron donors are available to them, so the postulated competition may have been between bacterial and algal photosynthesis by the same organism. Declining volcanic activity caused by falling temperatures within the earth may have caused a

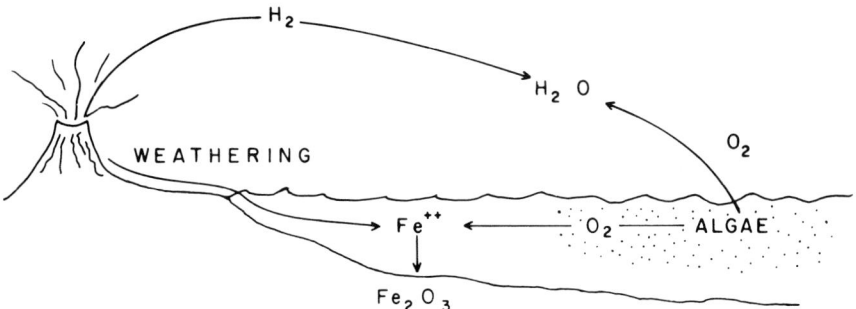

Fig. 8. The world remained anaerobic because the rate of production of oxygen by algae was less than the rate of production of reduced species by weathering and volcanoes.

progressive restriction in the rate of bacterial photosynthesis, permitting a corresponding growth in the rate of algal photosynthesis until oxygen began at last to accumulate in the atmosphere. An alternative suggestion, that biological productivity was restricted during the Archean by a lack of stable, shallow-water, marine environments, was made at this meeting by Knoll (1979).

Once it was being produced in excess, oxygen would have swept atmosphere and ocean free of reduced species, thereby eliminating photosynthetic bacteria as effective

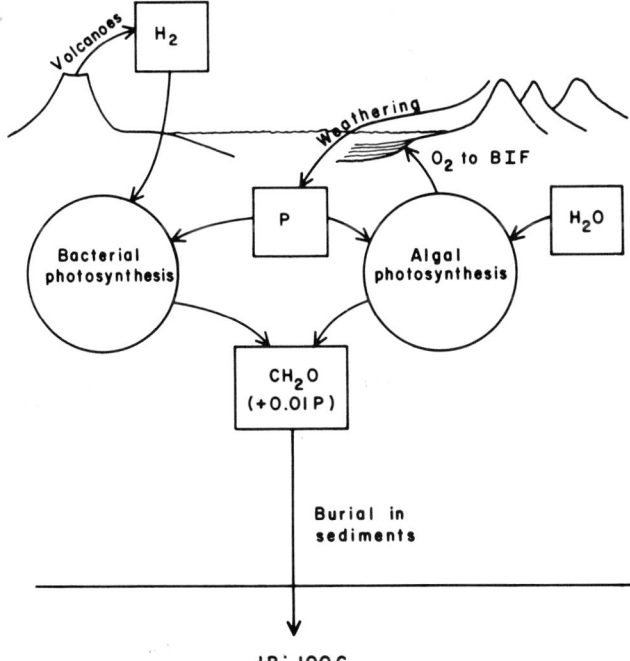

Fig. 9. The flow of phosphorus and hydrogen through the anaerobic ocean. The rate of bacterial photosynthesis was limited by the volcanic supply of hydrogen. The rate of algal photosynthesis was limited by the supply of phosphorus not consumed by photosynthetic bacteria (from Walker, 1978b).

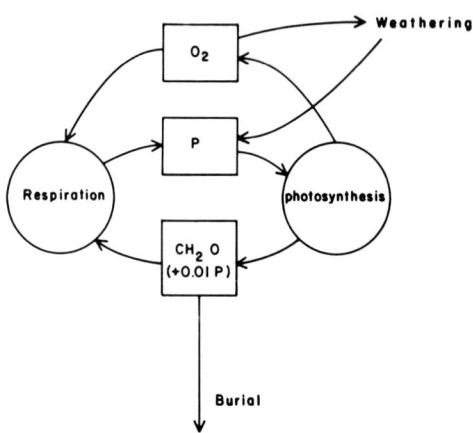

Fig. 10. The biogeochemical cycles that control atmospheric oxygen after the transition to an aerobic world (from Walker, 1978b). Phosphorus limits the rate of photosynthesis. The net oxygen source (excess of photosynthesis over respiration) is equal to the rate of burial of organic carbon (including some phosphorus), which in turn is equal to the rate of supply of phosphorus by weathering of rocks.

competitors. The rate of production of oxygen could have jumped, and the amount of oxygen in the atmosphere could have grown rapidly until some new process began to consume oxygen. This new process was presumably the process that consumes nearly all of the oxygen produced by algae and green plants today, namely aerobic respiration (Figure 10). Aerobic respiration was the last metabolic innovation to have a major impact on the atmosphere.

5. Conclusion

I have described a sequence of metabolic evolution that is broadly consistent with that outlined at this meeting by J. T. Staley as well as with the sequence of Margulis and Lovelock (1976) based on biochemical complexity and that of Schwartz and Dayhoff (1978) based on protein structure (cf. Margulis, 1970; Hall, 1971). Schwartz and Dayhoff suggest that aerobic respiration arose before algal photosynthesis, but it is hard to identify a source for the oxygen that would have made this possible (Walker, 1978a; see Towe, 1978, for dissent). I have, of course, had to ignore several important processes, including those involved in nitrogen metabolism.

The story, as I have described it, derives from the notion that organisms do not evolve without need or opportunity (Principle 3). Primitive microbes created needs by being greedy (Principle 4); opportunities, mainly new energy sources for heterotrophs, arose as side effects of the efforts of autotrophs to satisfy their needs. Prior to the origin of algal photosynthesis, biological productivity was limited by the volcanic source of hydrogen (which I use as a generic term to mean reduced gases). Each new development in the metabolic energy sources of autotrophs caused a reduction in the concentration of atmospheric hydrogen. Without this reduction there would have been no stimulus for the next development.

Although a picture emerges of coevolution of atmospheric composition and microbial metabolism there is a significant difference in the evolutionary processes. Metabolic capabilities evolved as a result of mutation and natural selection to overcome deficiencies in the environment; the atmosphere simply responded to what was done to it, although its response introduced new deficiencies in the environment leading to further evolution of metabolism. The direction of evolution was determined by the response of the atmosphere, but selection of innovations occurred on the biological level. Then, as now, the initiative rested with the biota.

Acknowledgements

This paper draws on material developed during the writing of a book (Walker, 1977). The many people who helped me in that endeavor are acknowledged in the Preface of that book. I am grateful to Lynn Margulis for answers to a number of questions. The National Astronomy and Ionosphere Center is operated by Cornell University under contract with the National Science Foundation.

References

Allaart, J. H.: 1976, in B. F. Windley (ed.), *The Early History of the Earth*, John Wiley and Sons, London, pp. 177–189.
Broecker, W. S.: 1971, *Quaternary Res.* 1, 188.
Broecker, W. S.: 1974, *Chemical Oceanography*, Harcourt Brace Jovanovich, New York.
Cloud, P. E.: 1968, *Science* 160, 729.
Cloud, P. E.: 1972, *Am. J. Sci.* 272, 537.
Cloud, P. E.: 1973, *Econ. Geol.* 68, 1135.
Cloud, P. E.: 1976, *Paleobiol.* 2, 351.
Cohen, Y., Padan, E., and Shilo, M.: 1975, *J. Bacteriol.* 123, 855.
Fridovich, I.: 1975, *Am. Sci.* 63, 54.
Fridovich, I.: 1978, *Science* 201, 875.
Garlick, S., Oren, A., and Padan, E.: 1977, *J. Bacteriol.* 129, 623.
Garrels, R. M., Lerman, A., and Mackenzie, F. T.: 1976, *Am. Sci.* 64, 306.
Hall, J. B.: 1971, *J. Theor. Biol.* 30, 429.
Holland, H. D.: 1973a, in *Proceedings of Symposium on Hydrogeochemistry and Biogeochemistry*, Vol. 1., Clarke Co., Washington, D.C., pp. 68–81.
Holland, H. D.: 1973b, *Econ. Geol.* 68, 1169.
Knoll, A. H.: 1979, *Origins of Life* 9, 313.
MacGregor, A. M.: 1927, *South African J. Sci.* 24, 155.
Margulis, L.: 1970, *Origin of Eukaryotic cells: Evidence and research implications for a theory of the origin of microbial, plant and animal cells on the Precambrian Earth*. Yale University Press.
Margulis, L. and Lovelock, J. E.: 1978, *Pure Appl. Geophys.* 116, 239.
Margulis, L., Walker, J. C. G., and Rambler, M.: 1976, *Nature* 264, 620.
Miller, S. L. and Orgel, L. E.: 1974, *The Origins of Life on Earth*, Prentice-Hall Inc., Englewood Cliffs, New Jersey.
Olson, J. M.: 1970, *Science* 168, 438.
Oren, A., Padan, E., and Avron, M.: 1977, *Proc. Nat. Acad. Sci. U.S.A.* 74, 2152.
Postgate, J. R.: 1968, in G. Nickless (ed.), *Inorganic Sulfur Chemistry*, Elsevier, New York, pp. 259–279.

Redfield, A. C.: 1958, *Am. Sci.* **46**, 205.
Redfield, A. C., Ketchum, B. H., and Richards, F. A.: 1963, in M. N. Hill (ed.), *The Sea*, Interscience, New York, Vol. 2, pp. 26–77.
Schidlowski, M.: 1979, *Origins of Life* **9**, 299.
Schwartz, R. M. and Dayhoff, M. O.: 1978, *Science* **199**, 395.
Stanier, R. Y., Douderoff, M., and Adelberg, E. A.: 1970, *The Microbial World*, 3rd edition, Prentice-Hall Inc., Englewood Cliffs, N.J.
Stephenson, M.: 1949, *Bacterial Metabolism*, 3rd edition, Longmans, Green and Co. Ltd., London, MIT Press Paperback Edition, 1966.
Towe, K. M.: 1978, *Nature* **274**, 657.
Van Valen, L.: 1971, *Science* **171**, 439.
Walker, J. C. G.: 1974, *Am. J. Sci.* **274**, 193.
Walker, J. C. G.: 1977, *Evolution of the Atmosphere*, Macmillan Publishing Co., New York.
Walker, J. C. G.: 1978a, *Pure Appl. Geophys.* **116**, 222.
Walker, J. C. G.: 1978b, *Pure Appl. Geophys.* **117**, 498.

ARCHEAN PHOTOAUTOTROPHY: SOME ALTERNATIVES AND LIMITS

ANDREW H. KNOLL

Department of Geology, Oberlin College, Oberlin, Ohio 44074, U.S.A.

Abstract. From the Archean geological record, one can infer that photoautotrophy evolved early in earth history; however, the nature of this photosynthesis — whether it was predominantly bacterial or cyanobacterial — is less clearly understood. General agreement that the earth's atmosphere did not become oxygen rich before the Early Proterozoic era places constraints on theories concerning more ancient biotas. Accommodating this limitation in various ways, different workers have hypothesized (1) that blue-green algae first evolved in the Early Proterozoic; (2) that oxygen producing proto-cyanobacteria existed in the Archean, but had no biochemical mechanism for coping with ambient O_2; and (3) that true cyanobacteria flourished in the Archean. but did not oxygenate the atmosphere because of high rates of oxygen consumption caused, in part, by the emanation of reduced gases from widespread Archean volcanoes.

Inversion of hypothesis three leads to another, as yet unexplored, alternative. It is possible that physiologically modern blue-green algae existed in Archean times, but had low productivity. Increased rates of primary production in the Early Proterozoic era resulted in the atmospheric transition documented in strata of this age. An answer to the question of why productivity should have changed from the Archean to the Proterozoic may lie in the differing tectonic frameworks of the two areas. The earliest evidence of widespread, stable, shallow marine platforms is found in Lower Proterozoic sedimentary sequnces. In such environments, productivity was, and is, high. In contrast, Archean shallow water environments are often characterized by rapid rates of clastic and pyroclastic influx — conditions that reduce rates of benthonic primary production.

This hypothesis suggests that the temporal correlation of major shifts in tectonic mode and atmospheric composition may not be fortuitous. It also suggests that sedimentary environments may have constituted a significant limit to the abundance and diversity of early life.

> Nothing is harder, yet nothing is more necessary, than to speak of certain things whose existence is neither demonstrable nor probable.
>
> Hermann Hesse

A *koan* is a riddle, usually in the form of a brief anecdote or aphorism, employed as a catechetical device in the discipline of Zen Buddhism. Initially the resolution of a *koan* may appear obvious, even trite; but as one continues to contemplate its subtleties, rolling them over in his mind, a variety of new and different interpretations suggest themselves. The riddle is an enigma. Each new insight leads down another path, and no two paths terminate at the same point. One example is the familiar paradox. "What is the sound of one hand clapping?" Another would be the somewhat more esoteric, "What is your face before your parents' birth?" The puzzle of Archean biology constitutes a *koan* of impressive complexity. Numerous lines of evidence, shaded by qualification or uncertainty, combine to reveal but a hazy picture of early evolution in which several widely varying scenarios seem possible.

It appears to be reasonably certain that life on earth did exist in Archean times. This statement can be justified, if for no other reason, by the recorded presence of stromatolites in sedimentary rocks up to 3000 m y. in age (MacGregor, 1941; Joliffe, 1955; Schopf et al., 1971; Bickle et al., 1975; Henderson, 1976; Mason and von Brunn, 1977). However, the nature of early organisms – their abundance and diversity – is less clearly understood. The reasons for this lie in the nature of the evidence itself. Archean stromatolites provide a good example. Most modern microbial mats are built principally by cyanobacteria; therefore, it might seem logical to assume that the oldest known mat-building communities also contained blue-green algae. Unfortunately, the problem is not so simple – bacteria can also build stromatolites. Flexibacteria weave coherent mats in hot spring environments of Yellowstone Park (Walter et al., 1972; Doemel and Brock, 1974), and it is possible that a bacterial community built the columnar stromatolites found in the 2000 m y. old Gunflint Iron Formation (Walter, 1972). Thus, the presence of stromatolites in Archean sedimentary sequences does not in and of itself constitute proof that oxygen eliminating photoautotrophs existed more than 2500 m.y. ago.

Other lines of evidence are similarly ambiguous. Microfossils preserved in Archean cherts (Muir and Grant, 1976; Knoll and Barghoorn, 1977; Dunlop et al., 1978) are morphologically similar to *Aphanocapsa*-type cyanobacteria, but they also resemble any one of a variety of bacteria, both photosynthetic and heterotrophic. Indeed, from the simple morphology of these bodies, it is difficult to distinguish them unequivocally from hypothesized prebiotic or protobiotic ancestors. The significance of the hematitic iron formation found in many Archean greenstone belts is likewise unclear. The fundamental question in this instance is, "Can significant concentrations of oxygen be produced abiologically by the photodissociation of water vapor in the upper atmosphere?" Opinion has been divided on this issue, but a recent contribution of K. M. Towe (1978) suggests that the answer is "yes" – at least in terms of early earth. Towe points out that molecular oxygen is required for one step in the biosynthesis of chlorophyll *a*. From this he draws the reasonable, in fact necessary, conclusion that oxygen tolerance preceded biological oxygen production. Small concentrations of oxygen were produced abiologically prior to the evolution of aerobic photosynthesis. Does this mean that O_2 bound in the hematite grains of Archean iron formation originated by the photodissociation of water molecules? Not necessarily. Nothing in Towe's biochemical arguments precludes the Archean appearance of physiologically modern cyanobacteria. The jury is effectively hung.

Carbon isotope ratios and the sedimentological distribution of organic matter constitute presumptive evidence for photosynthetic activity in early seas, but again, what kind of photosynthesis was involved, bacterial or cyanobacterial? And is it possible that abiotic processes of chemical evolution could have mimicked biological patterns of isotope fractionation and organic detritus distribution?

1. Limits Imposed by the Early Proterozoic Record

Earliest Proterozoic (and older) continental and deltaic deposits — for example, those of the South African Witwatersrand Supergroup or the lower part of the Huronian Supergroup in Ontario — ofter contain detrital uraninite and pyrite, but lack red beds. In contrast, red beds are widespread in younger Early Proterozoic sequences. In these rocks, it is the detrital uraninite and pyrite that are missing. It is widely accepted that this temporal change in the constitution of continental sedimentary rocks reflects a contemporaneous shift in the composition of the earth's atmosphere. Specifically, Early Proterozoic sedimentary strata record the initial appearance of free oxygen in the atmosphere (Cloud, 1968, 1973, 1974, 1976).

This interpretation has recently been challenged by Dimroth and Kimberley (1976) who believe that geologic evidence does not support the concept of a pre-Proterozoic oxygen-free atmosphere. It is impossible, in this brief paper, to summarize adequately their arguments, but it should be noted that some of the evidence documented by Dimroth and Kimberley, such as the absence of kerogen in littoral sandstones and limestones, does strongly suggest that at least *some* O_2 was present in the Archean atmosphere. (In the absence of at least a weakly oxidizing atmosphere, it is likely that small amounts of organic matter would have been preserved in these types of sediments.) A similar conclusion seems to be necessitated by the biochemical arguments of Towe (1978), as well as by the very presence of hematite-rich banded iron formation in Archean greenstone belts. Therefore, the critical question is not "*Was* oxygen present?"; it is "*How much* oxygen existed in the pre-Proterozoic atmosphere?" Dimroth and Kimberley maintain that essentially present-day concentrations of oxygen may have been present for the past 3800 m.y., but it is difficult to reconcile this hypothesis with the observed temporal shifts in the abundance of certain important types of sediments. A single example suffices to illustrate the point. If the oxygen content of the earth's atmosphere has not changed significantly during the past 3800 m.y., then it is difficult to explain why detrital uraninite, a widespread component of earliest Proterozoic sedimentary sequences, ceases to be important after that time. If, on the other hand, atmospheric oxygen did increase substantially during the Early Proterozoic, then the dearth of detrital uraninite in younger Precambrian strata can be explained readily (Cloud, 1973). The most reasonable conclusion is that the Archean atmosphere contained very small (<1% PAL, approximately the concentration at which detrital uraninite begins to oxidize rapidly) concentrations of O_2, and that these concentrations rose significantly during the Early Proterozoic era.

Curiously, it is this conclusion, based largely on evidence from Proterozoic rocks, that most strongly constrains hypotheses about Archean life. The way in which one explains the initial rise of atmospheric oxygen determines how one can interpret the carbon isotopic ratios and sedimentary distributions, the banded iron formation, the stromatolites, and the microfossils present in Archean sedimentary sequences. J. W. Schopf (1978) has hypothesized that the Early Proterozoic atmospheric transition reflects the initial appearance of the blue-green algae. In this view, green plant photosynthesis could have played no part in earlier microbial ecosystems. Thus, Archean stromatolites must be interpreted as the biosedimentary products of bacteria; carbon isotopic ratios and sedimentary distributions must be attributed to bacterial photosynthesis, to abiotic chemical syntheses, or to a combination of the two; and the hematite in early iron formations can only be related to the abiological production of O_2 via the photodissociation of water.

One alternative to this proposal is the scenario envisaged by Preston Cloud (1976). He also associates the Early Proterozoic rise of atmospheric O_2 concentrations with the evolution of physiologically modern cyanobacteria, but, unlike Schopf, he does not believe that the oxygen bound in Archean iron formations could have been produced abiologically. To accommodate the constraints of both atmospheric history and his interpretation of Archean oxygen production, Cloud has suggested that primitive cyanobacteria or protocyanobacteria flourished as early as 3800 m.y. B.P., but these *Uralgen* possesed no biochemical mechanism to cope with the oxygen released during photosynthesis and, thus, were restricted to environments in which ambient oxygen sinks could immediately combine with the toxic photosynthetic byproduct. Cloud regards ferrous iron in solution as a principle sink, and from this concludes that Archean iron formation is the product of primitive algal photosynthesis. According to this hypothesis, the exhaustion of available sinks early in the Proterozoic era created selective pressures that resulted in the evolution of oxygen tolerance and, ultimately, the rise of atmospheric oxygen. It follows from the tenets of this theory that the bulk of the organic carbon preserved in Archean strata must be a product of bacterial photosynthesis. Stromatolites may be viewed as either bacterial or blue-green algal.

Lynn Margulis and colleagues (1976) suggest that physiologically modern blue-green algae evolved early in earth history. In their discussion of atmospheric evolution, these authors note a simple but important feature of the terrestrial oxygen budget: free oxygen accumulates in the atmosphere when there is an excess of O_2 production *relative* to O_2 consumption. Thus, the apparent dearth of oxygen in the Archean atmosphere need not be ascribed to the absence or severe biochemical restriction of oxygen producing photoautotrophs; it could as easily have resulted from high rates of oxygen consumption. The abundance of lava and pyroclastic debris in Archean supracrustal sequences suggests that volcanism was extremely widespread in the early stages of crustal evolution. Margulis *et al.* hypothesize that it was the effusion of reduced gases from these volcanoes, coupled with such sinks as reduced minerals and organic matter, that kept Archean O_2 concentrations negligibly low, even though cyanobacteria were producing free oxygen. Waning volcanic intensity and oxygen saturation of surface minerals in the Early Proterozoic reduced rates of oxygen consumption to the point where they were lower than those of O_2 production. The result, an oxygen rich atmosphere. Under the terms of the argument advanced by Margulis *et al.*, stromatolites, microfossils, carbon isotope ratios and distributions, and banded iron formation can all be viewed as being predominantly the products of cyanobacterial photosynthetic activity. Schidlowski's (1976, 1977) theories regarding the carbon budget of the Archean earth would also be compatible with this type of scenario.

2. An Additional Alternative

William Smith (the essayist, not the father of stratigraphy) wrote in 1859 that:

if man was to think beyond what the senses had given him, he must first throw some wild guesswork into the air, and then, by comparing it bit by bit with nature, improve and shape it into truth.

It is in this spirit that I suggest one more way of thinking about Archean photoautotrophy and the rise of atmospheric oxygen. Bacteria do not arise by spontaneous generation, and neither do ideas. Thus, the proposal outlined below owes much to previous published hypotheses. It is perhaps most similar in outlook to that of Margulis *et al.* (1976), but differs in emphasis. Where Margulis *et al.* focused on oxygen *consumption*, I will concentrate on oxygen *production*.

> *The Hypothesis*: Communities of physiologically modern blue-green algae existed during the Archean era, but were characterized by low productivity. A significant increase in primary production during the Early Proterozoic era led to the oxygenation of the earth's atmosphere.

It is, of course, impossible to divine actual rates of photosynthesis for remote geological eras, but one can examine the environments available to early microbes and, by analogy to similar modern habitats, broadly categorize their potential for primary production. That is, one can apply a system of 'ecological uniformitarianism'. This approach has the same limitations as the physical uniformitarianism of Hutton (1795), but when applied to the Precambrian geological record, it yields results that are, at the very least, intriguing.

In 1943, F. J. Pettijohn compared Archean sedimentary patterns to those of Phanerozoic eugeosynclines. More recently, Dimroth and Kimberley (1976) have noted that "most Archean sedimentation apparently occurred on tectonically active, steep slopes surrounding volcanic piles". In these two observations, we find the essence of Archean sedimentary environments: rapid deposition of clastic and pyroclastic debris in tectonically active basins. Conglomerates, greywackes, and shales are common, while quartzose sandstones and carbonates are relatively infrequently encountered.

It might be assumed from Pettijohn's analogy that the surviving Archean sedimentary record consists largely of strata deposited in relatively deep water, but this is not the case. A growing body of evidence suggests that shallow water deposits are well represented among ancient clastic and pyroclastic accumulations. In a detailed sedimentological analysis of strata from the 3400 m y. old Onverwacht Group in the Barberton Mountain Land, South Africa, Lowe and Knauth (1977) documented the presence of such sedimentary features as cross-bedding, ripple marks, and desiccation-type cracks. Such evidence points to the conclusion that some of the Onverwacht sediments were deposited in shallow marine environments adjacent to subaerially exposed volcanic islands. Quartz arenites of the overlying Moodies Group, also located in the Barberton Mountain Land, contain numerous features associated with tidalities: cross stratification interference ripples, mudcracks, washouts and rill marks, and fining upward paleotidal range sequences (Eriksson, 1977; Klein and Ryer, 1978). Similar features characterize detrital deposits of the 3000 m.y. Pongola Supergroup, exposed south of the Barberton area (von Brunn and Hobday, 1977; Klein and Ryer, 1978). Coarse grained clastics characteristically occur in the marginal facies of Canadian Shield greenstone belts (Goodwin, 1973). Quiet water limestones and tidal flat mudstones do exist, and it is in these strata that the earth's oldest stromatolites are found. Such facies are rare however; perhaps half a dozen occurrences of stromatolites have been documented from an Archean geological record spanning more than 1000 million years.

Employing ecological uniformitarianism, one can speculate on the primary productivity of the Archean earth. Modern cyanobacterial mat communities have high rates of primary production - Bunt (1975) listed a value of 0.65-2.15 gC m^{-2} d^{-1} for intertidal blue-greens from Eniwetok, while Krumbein et al. (1977) recorded productivity rates as high as 12.0 gC m^{-2} d^{-1} from stromatolites from Solar Lake in the Sinai. It is reasonable to conclude that stromatolite building communities in Archean shallow, quiet water facies were also highly productive. But, as mentioned above, such habitats form only a very limited percentage of known early environments. More typical Archean shallow marine environments are characterized by high rates of clastic and pyroclastic influx, and in similar modern environments, benthonic productivity is relatively low - 0.08–0.53 gC m^{-2} d^{-1} for northern U.S. estuaries, 0.02 - 0.22 gC m^{-2} d^{-1} for tropical sediments, and 0.01 - 0.03 gC m^{-2} d^{-1} for Scottish shallow marine sediments (Bunt, 1975). Variation in methods of obtaining productivity data hinders attempts at detailed quantitative comparison of rates, but the data do establish that microbial mat communities in clastic-poor environments have higher rates of primary production per unit area than algae growing in regions of high clastic movement.

A significant problem arises in trying to assess the role of plankton in Archean primary production. Walker (in press) has offered the following intriguing statement:

It seems most unlikely that single-cell, freely floating organisms could have suvived in surface waters before the rise of oxygen and the development of the ozone screen. The real significance of ultraviolet light in Precambrian evolution may have been to prevent the development of planktonic forms of life prior to the rise of atmospheric oxygen.

The oldest assuredly planktonic microfossils are found in rocks of Early Proterozoic age (Hofmann, 1976; Knoll et al., 1978). The abundance of organic detritus in off-shore shale facies may suggest that the early ocean was populated by photosynthetic microorganisms, but this need not necessarily be the case. The organic matter contained in modern continental slope sediments is derived in part by transport from shallow water environments (Dow, 1977). Perhaps this also held true in Archean seas.

At present, prokaryotic photoautotrophs account for only a small fraction of the total primary production of open ocean ecosystems. In fact, only one taxon of cyanobacteria, the oscillatorian genus *Trichodesmium*, is commonly found among the oceanic plankton (Dawson, 1966). Does this situation reflect the competitive exclusion of blue-greens from pelagic environments by eukaryotic algae? Or was the oceanic habitat underexploited prior to the evolution of nucleated organisms? Here the limits of uniformitarian methodology became clear. However, if Walker is correct, then the problem of Archean productivity can be discussed completely in terms of benthonic communities. I will return to the question of planktonic productivity.

One can also infer from the preserved record of early sedimentation that the total area occupied by shallow marine environments was limited during the Archean era. Shallow water habitats were situated predominantly in narrow zones adjacent to volcanic arcs; extensive continental platform and miogeosynclinal shelf deposits are unknown in Archean supracrustal sequences.

Thus, while bearing in mind the obvious uncertainties involved, I suspect (but cannot prove) that whatever the dominant form of photoautotrophy in Archean seas, primary production was low relative to that of the succeeding Proterozoic era. The limited extent of shallow water marine environments and the common coupling of such regions to rapidly eroding land areas in active tectonic settings must have influenced the abundance and diversity of early biotas.

3. Early Proterozoic Environments

The geological differences between the Archean and Proterozoic are as fundamental as the biological dissimilarities between prokaryotic and eukaryotic cells. Table I summarizes the major features of Early Proterozoic sedimentary sequences. From this compilation emerges a picture of stable continental blocks flooded by epicontinental seas and bordered by geosynclines similar in size and developmental pattern to their Phanerozoic counterparts. This contrasts strongly with the granite-greenstone pattern of Archean rocks and represents the initial appearance of a more or less modern global sedimentary/tectonic regime in earth history (Windley, 1977). In terms of biological evolution, two features of this geological evolutionary transition seem particularly important:

1. The area available for colonization by benthic photoautotrophs must have increased substantially. Engel et al. (1974) have calculated that the ratio of quartzite plus carbonate to other sediments shifts from 1:100 in the Archean to about 1:5 in the Proterozoic. Windley (1977, p. 335) regards this as an approximate estimate of the increase in stable shelf and platform environments at this time, suggesting a twenty-fold increase in area.

2. The ecological rigor of these shallow water environments decreased significantly — again, one can contrast the thick sequences of carbonates that accumulated throughout wide stretches of the Early Proterozoic sea floor to the highly clastic habitats of the Archean.

Stromatolites first become widespread in rocks approximately 2200–2300 m.y. old (Schopf, 1975; Knoll, 1978). We know from preserved microfossils that at least some of

TABLE I
Lower proterozoic sedimentary sequences

Geologic unit	Location	Age (m.y.)	Tectonic setting	Iron formation ±	Red beds ±	Detrital uraninite ±	Carbonates	Stromatolites	Other notable features	Reference
1. Witwatersrand supergroup	South Africa	2600–2380	Continental basin	+	–[a]	+	–	–	Gold, thucolite	Pretorius (1974)
2. Huronian supergroup	Ontario, Canada	2300	Continental margin	–	+	+	+	–	Glaciogenic features	Young (1973a)
3. Montgomery lake beds	N.W.T., Canada	2300	Continental	–	–	+	–	–	–	Bell (1970)
4. Chibougamau and Otish mountain groups	Quebec, Canada	2300	Continental platform	–	–	+	–	–	Glaciogenic features	Young (1973b)
5. Jacobina group	Brazil	>2300	Continental	–	–	+	–	–	Gold	Cox (1967)
6. Mistassini group	Quebec, Canada	2200–1800	Platform or miogeosyncline	+	+	–	+	+	–	Chown and Caty (1973)
7. Animikie group	Lake Superior, U.S.A. & Canada	2000	Miogeosyncline	+	–	–	+	+	Coal	Goodwin (1956); Cannon (1973)
8. Circum-Ungava geosyncline	NE Canada	2000	Mio- and eugeo-syncline	+	+	–	+	+	Evaporites	Dimroth et al. (1970)
9. Coronation geosyncline	N.W.T., Canada	2200–1800	Mio- and eugeo-syncline, platform	–	+	–	+	+	–	Hoffman et al. (1970)
10. Hurwitz group	N.W.T., Canada	2300–1800	Geosyncline	+	+	–	+	+	–	Hoffman (1973); Bell (1970)
11. Sutton Lake group	Ontario, Canada	2000–1800	Miogeosyncline	+	?	–	+	+	–	Goodwin (1974)
12. Penrhyn and Piling grps.	Arctic Canada	2340–1950	Miogeosyncline	–	?	?	+	?	Metamorphosed	Heywood (1968)
13. Yavapai group	Arizona, U.S.A.	1820–1725	Variable pre- and synorogenic facies	+	–	–	–	–	Metamorphosed	King (1976)
14. Mt. Bruce supergroup	W. Australia	2200–1900	Miogeosyncline, platform	+	–	–	+	+	–	Daniels (1966); Horwitz and Smith (1977)
15. Nabberu group	W. Australia	2200–1900	Platform	+	?	–	+	+	–	Walter et al. (1976); Horwitz and Smith (1977)
16. Pine Creek Geosyncline	Australia	2200–1900	Geosyncline	–	+	–	+	+	–	Walpole et al. (1968)
17. Kimberley group	Australia	>1800	Continental margin or taphrogeo-syncline	–	+	–	–	–	All clastics	Canavan and Edwards (1938)

No.	Name	Location	Age (Ma)	Depositional setting						Notes	Reference
18.	Middleback group	S. Australia	>1800	?							Parkin (1969)
19.	Ketilidian supergroup	S. Greenland	>1800	Shallow basins, tectonically controlled						Coal	Bondesen (1970); Appel (1974)
20.	Ventersdorp supergroup	South Africa	2300	Volcanic basin with intercalated sediments	−	−	−	−	−	—	Winter (1963)
21.	Transvaal supergroup	South Africa, Botswana	2300–2100	Platform	+	−	+	+	+	—	Beukes (1973)
22.	Lomagundi group	Rhodesia	2300–2000	Platform	+	+	+	+	+	All clastics	Swift (1961)
23.	Waterberg system	South Africa	1900–1800	Non-marine in part	+	−	−	−	−	All clastics	Haughton (1969)
24.	Roraima formation	Guyana Shield, South America	1800	Continental basin	−	−	−	−	−		Grabert (1969)
25.	Minas series	Brazil	2700–1300	Miogeosyncline and platform		Metamorphosed and heavily weathered BIF					Dorr (1969, 1973)
26.	Carajas formation	Brazil	>1800	Platform	+	+[b]	−	−	+	Shungite (coal)	Dorr (1973)
27.	Karelian supergroup	Soviet Karelia and adjacent Finland	2500–2000	Miogeosyncline, platform	+	Heavily weathered BIF alternating with clastics			+		Salop (1977)
28.	Tampere group	Finland	>1900	Eugeosyncline	−	−	−	−	−	—	Salop (1977)
29.	Burzyan group	Southern Urals, U.S.S.R.	2470±500	Miogeosyncline, platform	+	−	+	+	+	Metamorphosed equivalents in N. Urals	Salop (1977)
30.	Krivoi Rog group and equivalents	Ukrainian S.S.R.	2200–1900	Miogeosyncline	+	−	−	+	+	'Coaly' phyllite	Salop (1977)
31.	Udokan group, Muya group	Baikal region U.S.S.R.	2600–1900	Platform, Miogeosyncline, eugeosyncline	+	+	−	+	+	—	Salop (1977)
32.	Charodokan to Kebetka formations	Aldan Shield, U.S.S.R.	2200–2000	Platform, aulacogen	−	+	−	+	+	—	Salop (1977)
33.	Khariton Laptev coast group and equivalents	Taymyr Belt, U.S.S.R.	>1900	Miogeosyncline, eugeosyncline	−	−	−	−	−	Metamorphosed clastics	Salop (1977)
34.	Liao Ho and Huto groups	China	2600–1860	Metamorphosed clastics and carbonates (some stromatolitic)							Salop (1977)
35.	Upper Dharwar and Aravalli groups	India	2600–2100	Geosyncline	+	−	−	+	+	—	Radhakrishna and Vasudev (1977); Salop (1977)

[a] The Witwatersrand red beds discussed by Dimroth and Kimberley (1976) are of secondary origin (Beukes, 1973).
[b] Detrital uraninite is found only in the basal beds of the Karelian sequence. Upper units contain red beds.

Fig. 1. Map showing the location of Lower Proterozoic sedimentary sequences of the world. Numbers indicating individual sequences refer to Table II.

these structures were built by essentially modern taxa of cyanobacteria (Hofmann, 1976; Nagy, 1978), and it is reasonable to suppose that *most* Early Proterozoic stromatolites are the product of blue-green algal activity. The sedimentary sequences containing these structures constitute the earliest record of extensive 'Proterozoic-type' platform carbonates. It is evident, then, that the blue-green algae rapidly took advantage of the appearance of new environments.

It is also in rock units of this age that we find the first evidence for the build-up of atmospheric oxygen. As mentioned earlier and documented in Table I, the youngest continental sediments containing detrital uraninite and pyrite are of earliest Proterozoic age (2600–2300 m.y.). Continental red beds appear soon afterward. In North America, for example, oxidized sediments fill grabens formed in the initial stages of the evolution of the Labrador Trough (Dimroth et al., 1970). Supratidal red beds, sometimes dotted by casts of halite and gypsum crystals, also occur near the base of the Lower Proterozoic sequence of the Belcher Islands (Jackson, 1960; Bell and Jackson, 1974). Other red beds can be found in the Hurwitz Group, the Mistassini Group, and in several horizons in the Coronation Geosyncline (see Table II for references), but the most important red beds in terms of fixing a date for the transition to an oxygen-rich environment are those present in the Huronian Supergroup of Ontario. This sequence contains detrital pyrite and uraninite in its lower half and oxidized sandstones in the upper. I see no reason to dispute S. M. Roscoe's hypothesis that the Huronian sedimentary deposits record the establishment of an oxygen-rich atmosphere (Frarey and Roscoe, 1970; Roscoe, 1973). Folding of the Huronian Supergroup 2200–2300 m y. ago fixes a minimum age for the atmospheric transition (Morris, 1977).

It has been common practice to assume that red beds did not appear until after the deposition of the world's great deposits of sedimentary iron formation, but it should be clear from the above arguments that this type of sediment was deposited contemporaneously with, and in some cases earlier than, the prodigious accumulations of Early Proterozoic iron formation. It is true that sedimentary sequences immediately overlying the great iron formations often contain thick red bed units — these deposits constitute the detritus eroded from continents uplifted during widespread orogenic episodes 2000–1800|m.y. ago. The red beds contained in these clastic blankets are conspicuous (the Roraima Formation, the Waterberg Group, the Athabasca and associated formations, and others); however, they are not the oldest deposits of their kind. The temporal overlap of red bed and iron formation deposition may reflect a brief but geologically resolvable period during which transport of ferrous iron in solution was possible in oceans lying beneath an increasingly aerobic atmosphere.

The important point to be gleaned from the above discussion is that the earliest evidence for extensive Proterozoic-type shallow carbonate seas, widespread stromatolitic communities, and oxygen-rich atmospheric conditions are all found in the same rock sequences. How does one explain this coincidence in the timing of fundamental geological, biological, and atmospheric changes? Following Cloud and Schopf, one can assume that important physiological innovations arose at or near the beginning of the Proterozoic era, *but one need not make this assumption*. Again applying the principle of ecological uniformitarianism one can assess the impact of the Archean/Proterozoic tectonic transition on a *pre-existing* cyanobacterial flora. The establishment of blue-green algal mat commumities over the vast new areas of shallow sea floor open to them would have had two important consequences:

1. Morphologic and perhaps biochemical diversity would have increased as cyanobacteria became more and more specialized to exploit smaller niches more successfully. Some evidence for this can be seen in the tremendous diversity of stromatolite types found in Early Proterozoic carbonates (Hoffman, 1976; Donaldson, 1976; Hofmann, 1976; Truswell and Eriksson, 1972, 1973).

2. Benthic productivity would have increased as a function both of increased area and of increased rates of production per unit area.

The extensive cratonization that ushered in the Proterozoic era also carries implications for increased upwelling and erosional run-off, the net effect of which would have been an increase in the productivity of oceanic plankton communities (Chamberlain and Marland, 1977). Increased rates of primary production, coupled with decreasing rates of oxygen consumption by reduced volcanic gases and progressively more oxidized surface sediments (Margulis et al., 1976), would have resulted in a rise in atmospheric oxygen concentrations. If Walker (in press) is correct that photosynthetic plankton could not have proliferated until the establishment of an ozone shield in the stratosphere, then this initiation of aerobic atmospheric conditions would have permitted the occupation of the vast ocean surface habitat. This, in turn, would also greatly increase the total primary production by O_2 eliminating microbes. In short, the superimposition of a change in the earth's tectonic framework upon a pre-existing cyanobacteria dominated biota would produce an Early Proterozoic record identical to that which really exists.

4. A Summary, Two Assumptions, and a Limit to Early Life

Consider two brief statements:

1. Organisms living in an aqueous environment evolved the capacity to ultilize water in the photosynthetic process early in the course of biological history.
2. The rise of atmospheric oxygen was an early consequence of the evolution of cyanobacterial photoautotrophy.

Both statements are logical, and both can be found woven into the fabric of many discussions of early evolution, but only one of them can be true. The hypothesis explored in this paper assumes the validity of sentence number one and presents environmental/ ecological reasons for suspecting that Archean blue-green algae existed for hundred of millions of years before they transformed the atmosphere into an oxygen-rich medium. A reiteration of major points of this arguments runs as follows:

1. While Precambrian organisms are most commonly treated as taxonomic entities, they can also be viewed from an ecologic perspective. The foci of such a treatment are the sedimentary environments available to early biotas.
2. Because of the limited extent and often rigorous nature of early shallow water environments, the primary production levels of an Archean cyanobacterial flora would have been low. Sinks for the O_2 produced by these organisms were present in great supply; thus, it is reasonable to expect that Archean atmospheric oxygen concentrations remained low – not because oxygen was not being produced, but because rates of production did not exceed rates of consumption.
3. The change in the earth's tectonic framework manifested in the Archean/Proterozoic boundary resulted in the initial appearance of widespread shallow platform and shelf environments. Blue-green algae rapidly covered those sea floors, resulting in the evolution of increasingly specialized biologically accommodated algal communities characterized by relatively high rates of photosynthetic production. For the first time in earth history O_2 production exceeded O_2 consumption, and the excess oxygen accumulated in the atmosphere.
4. The hypothesized production rates for Archean and Proterozoic biotas are consistent with what is known of the early geological record. In Proterozoic and Phanerozoic rock sequences, stromatolites, organic carbon isotope ratios, the distribution of carbon in sediments, and the oxygen bound in sedimentary iron formations are all routinely considered as products of cyanobacterial (and in younger sequences, higher algal) activity. This theory affords one the opportunity to consider Archean stromatolites, carbon

isotope ratios, carbon distribution patterns, and iron formations in the same light.

5. The hypothesis outlined in this paper allows one to consider the differences between Archean and Proterozoic biotas primarily in terms of ecological differences directly related to sedimentary/tectonic styles, thus avoiding the necessity of hypothesizing a major physiological advance coincident with the erathem boundary.

Two major assumptions underlie the arguments summarized in the preceding paragraph. The first is that the contribution of benthonic photoautotrophs to early ecosystems was significant with respect to total productivity. Recast in another fashion, this assumption suggests that if Archean planktonic microbes existed, their total primary production was low relative to the production of Proterozoic and younger plankters. The second assumption is that the preserved Archean sedimentary record is reasonably representative of the early earth as it actually existed. (In *The Origins of Species* (1859), Charles Darwin discussed at length the reasons why one should not expect to find forms transitional between species in the fossil record. Summarizing his arguments, Darwin states that anyone who does not believe that the geological record is really as he has described it "will rightly reject the whole theory". It is clear that my hypothesis also stands or falls on the interpretation of the rock record.)

Because we understand so little about the true nature of the Archean biosphere, it would be foolish of me to argue that the proposal advanced in this paper provides a better description of early biology than do other models. Quite honestly, I have not written this essay with that thought in mind. Rather, I have penned these thoughts in the belief that we need to be open to as many working hypotheses as possible. My hypothesis and all previously published theories have both strengths and weaknesses, and as new data become available we will be able to select the strong points and discard the weak. In this way it may indeed be possible to arrive eventually at some clear comprehension of the biological workings of the Archean earth.

A corollary of the ideas presented here ties directly into the theme of this symposium. I believe that no matter what type of phototrophs populated Archean environments, the availability and sedimentological nature of shallow marine habitats must have constituted important limits to the abundance and diversity of early life.

Acknowledgments

I thank H. D. Holland, L. Margulis, K. Niklas, C. Ponnamperuma, R. D. K. Thomas, and J. C. G. Walker for stimulating criticisms of an earlier draft of this paper. Preparation of this report was supported by the Department of Geology, Oberlin College.

References

Appel, P. U.: 1974, *Mineral. Deposita* 9, 75.
Bell, R. T.: 1970, *Geol. Surv. Can. Paper 70–40*, 159.
Bell, R. T., and Jackson, G. D.: 1974, *Can. J. Earth Sci.* 11, 722.

Beukes, N. J.: 1973, *Econ. Geol.* **68**, 960.
Bickle, M. J., Martin, A., and Nisbet, E. G.: 1975, *Earth Planet. Sci. Lett.* **27**, 155.
Bondesen, E.: 1970, *Medd. Grnl.* **185**, 210 pp.
Bunt, J. S.: 1975, *In:* Lieth, H. and Whittaker, R. H. (eds.) *Primary Productivity of the Biosphere,* Springer, Heidelberg, pp. 169–183.
Canavan, F., and Edwards, A. B.: 1938, *Australas. Inst. Min. Metall. Proc.* **110**, 59.
Cannon, W. F.: 1973, *Geol. Assoc. Can. Spec. Paper.* **12**, 251.
Chamberlain, W. M., and Marland, G.: 1977, *Nature* **265**, 135.
Chown, E. V., and Caty, J. L.: 1973, *Geol. Assoc. Can. Spec. Paper* **12**, 49.
Cloud, P. E.: 1968, *Science* **160**, 729.
Cloud, P. E.: 1972, *Am. J. Sci.* **272**, 537.
Cloud, P. E.: 1973, *Econ. Geol.* **68**, 1135.
Cloud, P. E.: 1974, *Am. Sci.* **62**, 54.
Cloud, P. E.: 1976, *Paleobiology* **2**, 351.
Cox, D. P.: 1967, *Econ. Geol.* **62**, 773.
Daniels, J. L.: 1966, *Australas. Inst. Min. Metall. Proc.* **219**, 17.
Darwin, C.: 1859, *The Origin of Species*, John Murray, London.
Dawson, E. Y.: 1966, *Marine Botany*, Holt, Rinehart, and Winston, New York.
Dimroth, E., Barager, W. R. A., Bergeron, R., and Jackson, G. D.: 1970, *Geol. Surv. Can. Paper* **70–40**, 45.
Dimroth, E., and Kimberley, M. M.: 1976, *Can. J. Earth Sci.* **13**, 1161.
Doemel, W. N., and Brock, T. D.: 1974, *Science* **184**, 1083.
Donaldson, J. A.: 1976, *In*: Walter, M. R. (ed.), *Stromatolites*, Elsevier, Amsterdam, 371.
Dorr, J. van N.: 1969, *U.S. Geol. Surv. Prof. Paper 641-A*, 110 pp.
Dorr, J. van N.: 1973, *Econ. Geol.* **68**, 1005.
Dow, W.: 1977, *A.A.P.G. Cont. Ed. Course Notes, Series 5*, 1.
Dunlop, H. S. R., Muir, M. D., Milne, V. A., and Groves, D. I.: 1978, *Nature* **274**, 676.
Engel, A. E. J., Itson, S. P., Engel, C. G., Stickney, D. M., and Cray, E. J., Jr.: 1974, *Geol. Soc. Amer. Bull.* **85**, 843.
Eriksson, K. A.: 1977, *Sed. Geology* **18**, 223.
Frarey, M. L., and Roscoe, S. M.: 1970, *Geol. Surv. Can. Paper* **70–40**, 143.
Goodwin, A. M.: 1956, *Econ. Geol.* **51**, 565.
Goodwin, A. M.: 1973, *Econ. Geol.* **68**, 915.
Goodwin, A. M.: 1974, *Am. J. Sci.* **274**, 987.
Grabert, H.: 1969, *Zentralb. Geol. Paläontol.* **3**, 523.
Haughton, S. H.: 1969, *The Geological History of Southern Africa.* Geol. Surv. S. Africa, Cape Town.
Henderson, J. B.: 1976, *Can. J. Earth Sci.* **12**, 1619.
Heywood, W. W.: 1968, *Geol. Surv. Can. Paper* **66–40**, 20 pp.
Hoffman, P. F.: 1973, *Roy. Soc. London Phil. Trans., Ser. A.* **273**, 547.
Hoffman, P. F.: 1976, *In*: Walter, M. R. (ed.) *Stromatolites*, Elsevier, Amsterdam, pp. 599–612.
Hoffman, P. F., Fraser, J. A., and McGlynn, J. C.: 1970, *Geol. Surv. Can. Paper* **70–40**, 201.
Hofmann, H. J.: 1976, *J. Paleontol.* **50**, 1040.
Horwitz, R. C., and Smith, R. E.: 1977, *Precambrian Res.* **6**, 293.
Hutton, J.: 1795, *Theory of the Earth with Proofs and Illustrations.* Edinburgh.
Jackson, G. D.: 1960, *Geol. Surv. Can. Paper* **60–20**, 13 pp.
Joliffe, A. W.: 1955, *Econ. Geol.* **50**, 373.
King, P. B.: 1976, *U.S. Geol. Surv. Prof. Paper* **902**, 85 pp.
Klein, G. deV., and Ryer, T. A.: 1978, *Geol. Soc. Am. Bull.* **89**, 1050.
Knoll, A. H.: 1978, *Nature* **276**, 701.
Knoll, A. H., and Barghoorn, E. S.: 1977, *Science* **198**, 396.
Knoll, A. H., Barghoorn, E. S. and Awramik, S. M.: 1978, *J. Paleontol.* **52**, 976.
Krumbein, W. E., Cohen, Y., and Shilo, M.: 1977, *Limnol. Oceanogr.* **22**, 635.
Lowe, D. R., and Knauth, L. P.: 1977, *J. Geol.* **85**, 699.
MacGregor, A. M.: 1941, *Trans. Geol. Soc. S. Africa* **43**, 9.
Margulis, L., Walker, J. C. G., and Rambler, M.: 1976, *Nature* **264**, 620.

Mason, T. R., and von Brunn, V.: 1977, *Nature* **266**, 47.
Morris, W. A.: 1977, *Geology* **5**, 137.
Muir, M. D., and Grant, P. R.: 1976, *In*: Windley, B. F. (ed.) *The Early History of the Earth*, Wiley, New York, pp. 595–604.
Nagy, L. A.: 1978, *J. Paleontol.* **52**, 141.
Parkin, L. W. (ed.): 1969, *Handbook of South Australian Geology*. S. Austral. Geol. Surv., Adelaide, 268 pp.
Pettijohn, F. J.: 1943, *Geol. Soc. Am. Bull.* **54**, 925.
Pretorius, D. A.: 1974, *Univ. Witswatersrand Econ. Geol. Res. Unit. Inf. Circ.* **86**, 50 pp.
Radhakrishna, B. P., and Vasudev, U. N.: 1977, *J. Geol. Soc. India* **18**, 525.
Roscoe, S. M.: 1973, *Geol. Assoc. Can. Spec. Paper* **12**, 31.
Salop, L. J.: 1977, *Precambrian of the Northern Hemisphere*, Elsevier, Amsterdam, 378 pp.
Schidlowski, M.: 1976, *In*: Windley, B. F. (ed.) *The Early History of the Earth*, Wiley, New York, pp. 525–534.
Schidlowski, M., and Eichmann, R.: 1977, *In*: Ponnamperuma, C. P. (ed.) *Chemical Evolution of the Early Precambrian*, Academic Press, New York, pp. 87–99.
Schopf, J. W.: 1975, *Ann. Rev. Earth Planet. Sci.* **3**, 213.
Schopf, J. W.: 1978, *Sci. Am.* **239** (3), 110.
Schopf, J. W., Oehler, D. Z., Horodyski, R. J., and Kvenvolden, K. A.: 1971, *J. Paleontol.* **45**, 477.
Smith, W.: 1859, *Quoted in*: Eiseley, L.: 1961, *Darwin's Century*, Anchor, Garden City, New York, p. 117.
Swift, W. H. (ed.): 1961, *An Outline of the Geology of Southern Rhodesia. Bull. Geol. Surv. S. Rhodesia* **50**.
Towe, K. M.: 1978, *Nature* **274**, 657.
Truswell, J. F., and Eriksson, K. A.: 1972, *Trans. Geol. Soc. S. Africa* **75**, 99.
Truswell, J. F.: 1973, *Sed. Geol.* **10**, 1.
Von Brunn, V., and Hobday, D. K.: 1976, *J. Sed. Petrol.* **46**, 670.
Walpole, B. P., Dunn, P. R., and Randal, M. A.: 1968, *Austral. BMR Bull.* **82**, 169 pp.
Walker, J. C. G.: *Pure Applied Geophys.* **116**, 222.
Walter, M. R.: 1972, *Econ. Geol.* **67**, 965.
Walter, M. R., Bauld, J., and Brock, T. D.: 1972, *Science* **178**, 402.
Walter, M. R., Goode, A. D. T., and Hall, W. D. M.: 1976, *Nature* **261**, 221.
Windley, B. F.: 1977, *The Evolving Continents*, Wiley, New York, 385 pp.
Winter, H. de la R.: 1963, *Trans. Geol. Soc. S. Afr.* **66**, 115.
Young, G. M. (ed.): 1973a, *Geol. Assoc. Can. Spec. Paper* **12**, 271 pp.
Young, G. M.: 1973b, *Geol. Assoc. Can. Spec. Paper* **12**, 97.

SULPHUR ISOTOPE RATIOS IN LATE AND EARLY PRECAMBRIAN SEDIMENTS AND THEIR IMPLICATIONS REGARDING EARLY ENVIRONMENTS AND EARLY LIFE*

H. G. THODE

Department of Chemistry, McMaster University, Hamilton, Ontario, Canada

Abstract. The stable isotopes of sulphur are fractionated in equilibrium and unidirectional processes in the earth's crust and biosphere. By far the most important of these processes occur in the biological sulphur cycle characterized by the activity of sulphur oxidizing and reducing microbiota. In particular, the dissimilatory reduction of sulphate to hydrogen sulphide by anaerobic bacteria leads to isotope effects of from 0 to $\sim 60^{0}/_{00}$, the magnitude of the effect depending largely on metabolic rates. Actual isotope ratio ($\delta^{34}S$) patterns in sediments depends, therefore, on environmental conditions and the nature of sulphate reservoirs during reduction. Sulphur isotope ratios can and have been used to trace environmental conditions, sources, and modes of formation of certain Phanerozoic deposits.

These studies which have been extended to late and early Precambrian sediments provide a potential source of information about very early sediment deposition environments and early life. Recent carbon and sulphur isotope data for the low grade metamorphosed banded iron-formations of the Michipicoten area in Ontario (2.7 b.y. old) provide strong evidence for the existence of autotrophic organisms and reducing bacteria in late Archean times.

Sulphur isotope ratios ($\delta^{34}S$) have now been obtained for samples from the Isua area of West Greenland. The $\delta^{34}S$ of the Isua sediments (3.7 b.y. old), including the various facies of the banded iron-formations, have a very narrow spread with their mean close to zero $^{0}/_{00}$ C.D.T. (0.45 ± 0.5). This comes extremely close to the respective means yielded by the Isua tuffaceous amphibolites (+0.3 ± $0.9^{0}/_{00}$) and by the somewhat younger, 3.1 to 3.7 x 10^{9} yr, basaltic Ameralik dykes of the region (+0.6 ± $1.1^{0}/_{00}$).

These results indicate a complete absence of isotopic evidence for 'sulphate reducers' in the Isua sediments (early Archean) in contrast to the banded iron-formations of the late Archean, where $\delta^{34}S$ varies from -2- to $+20^{0}/_{00}$.

1. Introduction

The isotopes of sulphur may be fractionated in unidirectional and or equilibrium processes involving the various oxidation states of sulphur both in geological and biological processes. The biological sulphur cycle characterized by the activity of sulphur oxidizing and reducing bacteria, however, account for the bulk of contemporary turnover of sulphur. In this cycle, the dominant mechanism of sulphur isotope fractionation is the dissimilatory reduction of sulphate to sulphide by the bacterium Desulpho vibrio desulphuricans.

---

* Contribution No. 90 of the McMaster Isotopic, Nuclear and Geochemical Studies Group.

2. Simple Process Isotope Effect

In this process, occurring in reducing sediments at the bottom of the sea, the hydrogen sulphide produced is depleted in the heavy isotope of sulphur ^{34}S by from 0 to $\sim 50‰$ with respect to the source sulphate pool (Thode et al., 1951; Harrison and Those, 1958; Kaplan and Rittenberg, 1964; Kemp and Thode, 1968). The magnitude of this isotope effect depends on conditions of temperature, nutrient concentrations, reducing conditions, etc., in general decreasing with increasing metabolic rate. Isotope studies of recent marine sediments suggest that less than ideal conditions prevail in natural marine sediments and that metabolic rates are in general relatively slow favouring the higher isotope effects ($25‰-60‰$) (Kaplan et al., 1963; Vinogradov et al., 1962). Studies of sulphur isotope ratio ($\delta^{34}S$) distribution patterns in sediments can therefore give us information as to the nature of the environment during sediment deposition and evidence of biological activity.

Sulphur isotope ratios are expressed in terms $\delta^{34}S$ defined as follows:

$$\delta^{34}S, ‰ = \frac{(^{34}S/^{32}S)\text{ sample} - (^{34}S/^{32}S)\text{ standard}}{(^{34}S/^{32}S)\text{ standard}} \times 1000.$$

The standard ratio used is that for the troilite sulphur of the Canyon Diablo meteorite. There is considerable evidence to show that this meteoritic ratio $\delta^{34}S = 0$ on the above meteorite scale, is close to the primordial ratio of the sulphur isotopes for our solar system. For example, the average sulphur isotope ratio for basic sill sulphur of deep seated origin is close to that of the standard meteorite or $\delta^{34}S \cong 0$ (Shima et al., 1963).

The sulphur isotope effect in the bacterial reduction of sulphate is then the difference between two sulphur isotope ratios or $\delta^{34}S$ values and is given by:

Isotope effect $(\delta, ‰) = (\alpha - 1)1000 = \delta^{34}S, (SO_4^=) - \delta^{34}S, (S^=)$. Where α is the isotope separation factor and $\delta^{34}S, (SO_4^=)$ and $\delta^{34}S, (S^=)$ are the corresponding isotope ratios of the source sulphate and product hydrogen sulphide. In laboratory experiments, this isotope effect $\delta, ‰$ may be determined directly since corresponding $\delta^{34}S, (SO_4^=)$ and $\delta^{34}S, (S^=)$ can both be measured.

The isotope effect prevailing in natural settings, such as recent marine sediments, may also be determined in some cases. $\delta^{34}S, (S^=)$ for the reduced sulphur in the sediments can be measured, and $\delta^{34}S, (SO_4^=)$ may be equated to $\delta^{34}S$, (ocean $SO_4^=$) ($\sim +20‰$) assuming sulphate reduction in an open system, in equilibrium with the large ocean sulphate reservoir. However, some sulphate may be reduced in a closed system, in the muds after burial. Complete reduction of this residual sulphate cannot lead to any net change in isotope ratio. Thus the apparent isotope effect will be reduced in proportion to the fractionation of sulphate reduced after burial. This may be estimated, and the isotope effect prevailing in a natural environment may be determined from sulphur concentrations and sulphur isotope ratios as a function of depth in recent marine sediments.

The proportions of sulphate reduced in open and closed systems would seem to depend in part on sedimentation rates. In the deep Black Sea environments where sedimentation rates are only ~0.1 mm yr^{-1} the apparent isotope effects are near the maximum attainable or ~50‰, indicating little or no reduction of sulphate after burial (Monster and Thode, 1978). However, in recent sediments off the coast of Venezuela, where sedimentation rates are high, ~25 mm yr^{-1}, the apparent isotope effect is about $\frac{1}{3}$ of the maximum or ~15‰, suggesting that a considerable amount perhaps ~$\frac{2}{3}$, of the sulphate is reduced after burial (Thode et al.,1960).

In the case of ancient Phanerozoic sedimentary rocks of marine origin $\delta^{34}S$, $(S^=)$ reduced sulphur can again be measured. However, $\delta^{34}S$ of the pool of sulphate from which the reduced sulphur was derived is not obtainable. Thus the simple process isotope effect prevailing in sulphate reduction at the time of sediment deposition cannot be determined directly.

$\delta^{34}S$ of the sulphate in the present day oceans is uniform in breadth and depth and has been essentially constant at ~+20‰ over the past 20 m yr (Thode et al., 1961). However, we know from studies of sulphur isotope ratios in marine evaporites ($CaSO_4$) associated with ancient oceans that since Cambrian time it has varied by at least 20‰ (Thode and Monster, 1965; Holser and Kaplan, 1966). Thus, on a large time scale the oceans provide a finite reservoir of sulphate which is subject to batch fractionation in the reduction sulphate to sulphide by 'sulphate reducers' (Rees, 1969).

Furthermore it is known that ancient inland seas partially cut off from oceans or restricted sulphate reservoirs have had quite different $\delta^{34}S$, $(SO_4^=)$ values from their contemporaneous oceans and that their δ values have varied markedly over relative short periods of time. Batch process fractionation in the bacterial reduction of sulphate is the dominant factor in these changes in $\delta^{34}S$ (Thode and Monster, 1965).

3. Batch Process Fractionation

Since in the bacterial reduction of sulphate, the sulphide produced and precipitated in the sediments is depleted in the heavy isotope ^{34}S, the residual sulphate in a closed or restricted basin will become increasingly enriched in this heavy component, as in a Rayleigh distillation process. The magnitude of this batch process isotope enrichment depends on the fraction of the sulphate in the restricted basin which has been reduced, as well as on the simple isotope effect in the bacterial reduction process (Thode, 1970). Thus, high $\delta^{34}S$, $(SO_4^=)$ enrichments in a marine basin are possible as the sulphate becomes reduced. Intermittent flow of fresh sulphate into the basin (sea transgressions) would then produce large swings in $\delta^{34}S$, $(SO_4^=)$ values.

The variability and high positive $\delta^{34}S$, $(SO_4^=)$ values of many Phanerozoic evaporite deposits, gypsum and anhydrite, are due to this type of batch or Rayleigh fractionation involving 'sulphate reducers' in restricted basins. Variations in $\delta^{34}S$, $(S^=)$ of the sulphides deposited in the basin sediments with time will, of course, parallel the swings of $\delta^{34}S$, $(SO_4^=)$ of the basin sulphates the corresponding δ values being displaced by the sulphate reduction isotope effect.

Burnie et al. (1972) have reported this kind of $\delta^{34}S$, $(S^=)$ distribution pattern in the late Precambrian Nunesuch Shale (1.1 b.y.) of Michigan and have discussed bacterial reduction models to explain them. Thode et al. (1962) and Chukhrov et al. (1970) have reported δ values for sulphides and sulphates in Precambrian sedimentary rocks ranging up to 50‰. These large variations and high +δ values have been considered as evidence for dissimilatory 'sulphate reducers' in these rocks (1.8 − 2.5 b.y. old).

4. Archean Sediments

Since 'sulphate reducers' require organic matter with suitable hydrogen donors and sulphate in their metabolism, it is generally assumed that their beginnings in Archean times was proceeded by sulphur oxidizers and photosynthetic bacteria. Some build-up of oxygen in the atmosphere may also have been required.

There is considerable evidence to suggest the presence of only trace amounts of oxygen in the late to middle Archean. The models developed to explain the formation of the Archean banded iron-formations are based on an oxygen deficient atmosphere (James, 1954; Goodwin, 1964; Holland, 1973). Also the rare occurrences of sulphates and evaporites in the Archean sedimentary rocks suggests limited concentrations of sulphate in the Archean oceans. However, recently several barite deposits have been reported in Archean sediments (Chukhrov et al., 1970; Perry et al., 1971; Vinogradov et al., 1976).

Perry et al. (1971) and Heinrichs and Reimer (1977) have given fairly convincing evidence that the extensive beds of barite alternating with the green cherts, shales and conglomerates of the Fig Tree Swaziland sequence in South Africa, are indeed primary sediments deposited in close association with the Fig Tree group $\sim 3.2 \times 10^9$ yr ago. Perry et al. (1971) suggested the oxidation of sulphide to sulphate by chromatium sp. (Sulphur oxidizing photosynthetic bacteria) as the sulphate source of the Fig Tree barites rather than sulphate from the weathering of sulphides in an oxygenated atmosphere. The small displacement of $\delta^{34}S$ of these barites from zero, $\sim 3‰$, is consistent with this proposal since chromatium sp. produce little if any sulphur isotope fractionation (Kaplan and Rittenberg, 1964).

In regard to the bacterial reduction of sulphate to sulphide in the presence of low concentrations of sulphate, Harrison and Thode (1958) have demonstrated that in the concentration range below 10^{-3} molar, the isotope effect falls off to zero and at 10^{-5} molar is reversed slightly in direction ($\sim 2‰$ enrichment of ^{34}S in the sulphide).

In any case, the pattern of sulphur isotope distribution $\delta^{34}S, (S^=)$ in Archean sediments may be compared with characteristic patterns found in younger rocks where bacterial reduction is known to have taken place. A highly $\delta^{34}S, (S^=)$ distribution pattern both on a micro and macro scale within the sedimentary rock formations will provide strong evidence for the existence of sulphate reducing bacteria at the time of sediment deposition.

5. $\delta^{34}S$ Distribution Patterns in Michipicoten and Woman River Iron-Formations

As in all stable isotope tracer studies in Precambrian sediments, one must be concerned about the possibility of alteration of isotope ratios since their deposition due to mineral additions, mineral replacement, isotope fractionation during sediment diagenesis and sediment metamorphism.

The Archean Banded Iron-Formations of the Michipicoten and Woman River areas of Ontario, 2.75×10^9 yr old, (Goldich, 1973) are good examples of well-preserved low grade metamorphic rocks ($\sim$200 °C) and provide an unusual opportunity to study isotope distribution patterns in depth and to test the various models for their origin and the extent of biological activity in their derivation (Goodwin, 1964; Goodwin et al., 1976).

These iron-formations in the Michipicoten basin are about 50 x 70 miles in extent up to $\sim$1500 feet thick underlain and overlain by felsic volcanic and mafic volcanic rocks respectively (Goodwin, 1964, 1973). They display a distribution pattern of oxide (magnetite), carbonate (siderite), and sulphide (pyrite and pyrrhotite) facies based on the dominant iron minerals present. The three iron-facies each composed of interbedded chert, broadly transitional one to the other, across the area serve as sensitive indicators of local sedimentary environments during periods of chemical deposition, lithification and consolidation. This iron-facies pattern common to many Animikie and Archean banded iron-formations are interpreted to reflect an inclined paleoslope on which oxide facies are deposited comparatively near shore in a mildly oxidizing environment, carbonate-facies in somewhat deeper waters mildly reducing and sulphide-facies in still deeper waters markedly

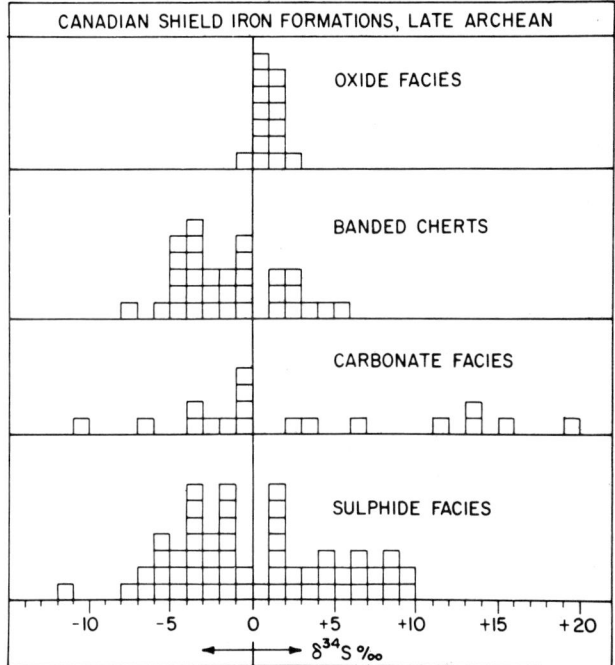

Fig. 1. $\delta^{34}S\%_{oo}$ C.D.T. distribution patterns of sulphides in drill core samples covering 2000 km of Michipicoten, Woman River and Bending Lake iron-formation sediments in the Canadian Shield (2.75×10^9 yr old) (Goldich, 1973). Oxide facies (magnetite) from Bending Lake formation, carbonate facies (siderite), sulphide facies (pyrite and pyrrhotite) and banded cherts from the Michipicoten and Woman River iron-formations (same Archean basin) (Goodwin et al., 1976).

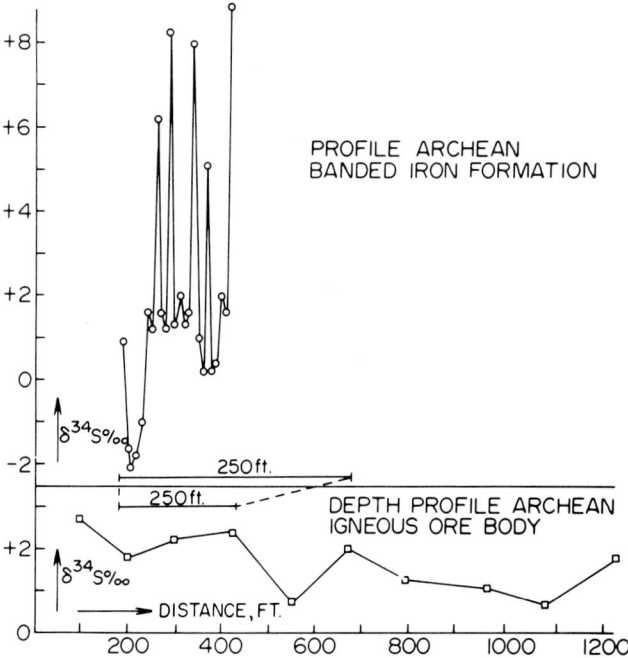

Fig. 2. A typical $\delta^{34}S\%_{oo}$ C.D.T. profile of Archean (Michipicoten) banded iron-formation compared to that of an Archean igneous ore body (Yellow Knife district Canadian Shield) of about the same age (Goodwin et al., 1976; Wanless et al., 1960).

reducing environment. The bacterial reduction of sulphate in the strongly reducing environment was proposed as the main source of the H_2S in the formation of the sulphide-facies of the Animikie and Archean banded iron-formations (James, 1954; Goodwin, 1964; Garrels et al., 1973; Holland, 1973). Some ten Archean basins in the Canadian Shield have been identified similar to that of the Michipicoten and Woman River basin, (outlined by the common shore to depth facies pattern of the iron-formations). Two of these basins including the Michipicoten have been established as completely closed (Goodwin, 1964).

$\delta^{34}S, (S^=)$ distribution patterns for the banded chert, siderite, sulphide facies of the Archean Michipicoten and Woman River iron-formations reported by Goodwin et al. (1976), and for the oxide facies of the Archean Bending Lake formation observed by Thode et al. (1979), are shown in Figure 1. In Figure 2, an example of a $\delta^{34}S$ profile of an Archean banded iron-formation section is compared with that of an Archean igneous ore body. The chert siderite and sulphide facies of the iron-formations considered to be chemical precipitates formed in mildly to strongly reducing environments all have similar $\delta^{34}S, (S^=)$ distribution patterns. In total $\delta^{34}S$ values vary from -10.5 to $10.7\%_{oo}$, have a mean value of $-0.7\%_{oo}$ and a range of $21.2\%_{oo}$. Isotopic analysis of an additional drill core sample have extended the range of $\delta^{34}S$ values to $30\%_{oo}$ (Thode and Monster, 1979).

6. The Bacterial Reduction Model

This high variability in $\delta^{34}S, (S^=)$ exhibited throughout the sections of the siderite, sulphide facies, both on a macro and micro scale is characteristic of sedimentary sulphides formed by bacterial reduction of sulphate in a closed or restricted basin where batch process fractionation can occur. However, the range in $\delta^{34}S$ values is somewhat less than that found in some younger biogenic sediments. For example, sulphides in the Nunesuch Shales in Michigan late Precambrian in age (1.1 b.y.) known to be biogenic in origin have $\delta^{34}S, (S^=)$ ranges of from $35°/_{oo}$ to $48°/_{oo}$ (Burnie et al., 1972). However, differences in $\delta^{34}S, (S^=)$ ranges and distribution patterns in ancient biogenic sediments are to be expected in view of the many different factors involved. As pointed out above, if bacterial reduction takes place in a restricted basin of limited sulphate concentration, then $\delta^{34}S$ of the source sulphate will increase with the fraction reduced until fresh sulphate is introduced. Therefore, ranges in $\delta^{34}S$ dependent on batch process isotope enrichment are controlled by flow rates of fresh sulphate into the basin. The banded nature of the iron-formations suggests also an intermittent flow of ferrous iron into the active part of the basin where iron oxide, siderite and sulphides are deposited. In the absence of ferrous iron, H_2S would build up in the deeper anaerobic waters of high biological activity. This cycle of H_2S accumulation and then precipitation as iron sulphide would reduce the amplitude of the $\delta^{34}S$ swings of the Rayleigh fractionation process.

A smaller simple isotope effect in the bacterial reduction process itself such as may have prevailed in Archean times under particular conditions could also account in part for the smaller range of $\delta^{34}S$ values found. This isotope effect varies with metabolic rates, and local changes in environmental conditions can give rise to $\delta^{34}S, (S^=)$ variations within a single sediment sample. Thus, the lack of homogeneity in isotope ratio within a single specimen of sedimentary rock is characteristic of sedimentary sulphides of bacterial origin. This kind of small scale variability in $\delta^{34}S$ is exhibited in the siderite-sulphide samples of the B.I.F. as well as the large scale variability shown in Figure 1.

The oxide-facies (magnetite) from the equivalent Bending Lake Archean basin also ~2.75 b.y. old, considered to have been formed by chemical precipitation in a local oxidizing environment near shore, exhibits a very narrow range of $\delta^{34}S$ values close to zero not unlike primitive sulphur of igneous origin (Figure 1). In contrast to the siderite-sulphide facies distribution data, there is little evidence of isotope fractionation due to 'sulphate reducers' in this local oxidizing environment in which iron oxide was deposited.

The sulphide concentrations in the oxide facies samples are low 0 to 1.5% as compared to 2 to 90% for the siderite-sulphide facies. The results suggest possibly a submarine volcanic exhalation source of H_2S with the primordial ratio ($\delta^{34}S \cong 0$) for the precipitation of the small quantities of sulphide in the oxide facies. There could be a small component of this primary sulphur in the banded chert, siderite and sulphide facies as well. Samples of these facies with low sulphur contents could account for the high preponderance of $\delta^{34}S$ values $\cong 0$, see Figure 1.

Organic matter is another source of reduced sulphur that could lead to relatively small quantities of sulphides in sediments. Since no isotope effect has been detected in the plant metabolism of sulphate (Ishii, 1953; Kaplan et al., 1963) $\delta^{34}S$, $(S^=)$ would depend on that of the source sulphate.

Goodwin et al. (1976) have ruled out volcanic gases (submarine exhalations) as a more direct source of H_2S for the precipitation of the large quantities of sulphides found in the siderite-sulphide facies of the B.I.Fs. It is argued that the quantities of H_2S available during the quiet depositional periods would not be sufficient and that H_2S derived from volcanic gases would almost certainly have δ values essentially equal to zero (Goodwin et al., 1976). It is concluded that only the bacterial reduction model of sea water sulphate to H_2S in a restricted Archean basin can explain the large sulphide deposits of the Michipicoten and Woman River iron-formations and their highly variable sulphur isotope distribution patterns.

Carbon isotope ratios, $(\delta^{13}C)$, ranging from $-20\permil$ to $-28\permil$ for the graphitic carbon in the cherts and $\delta^{13}C$ values ranging from $-1\permil$ to $+2.5\permil$ for the carbonate carbon in massive siderite of the Michipicoten and Woman River iron-formation, also reported by Goodwin et al. (1976), are not inconsistent with the above conclusions. In this regard, Karkhanish of the University of Western Ontario (private communication) has observed evidence of algae mat colonies in a chert unit of the Michipicoten iron-formation.

7. $\delta^{34}S$ Distribution Patterns in Isua Sediments Early Archean

The banded iron-formations of the Isua area of West Greenland are the oldest sedimentary rocks known to date, 3.7 b.y. old, (Moorbath et al., 1973). Recently Monster et al. (1979), measured the percent sulphur and sulphur isotope ratios $\delta^{34}S$ in samples collected from these early Archean sediments to further narrow down the time of the rise of 'sulphate reducers' during the Precambrian.

Different facies of the iron-formation have been observed within the supracrustals. Most common is the oxide facies. Less frequent are 'carbonate' facies, 'silicate' facies and 'sulphide' facies. Oxide facies occur throughout the whole belt (~30 km long and 3 km wide), but mostly in the northeastern part, whereas the most reducing facies seems to be most abundant in the western part of the belt. Based on field and textural evidence, it can be demonstrated that the sulphides were precipitated contemporaneously with the precipitation of the iron, silica and carbonate in the banded iron-formation, and contemporaneously with deposition of the basaltic tuffs (Appel, 1978).

The sulphides therefore appear to be syngenetic. Metamosphism might have caused slight shifts in sulphur isotopic ratios in coexisting sulphides, the extent of this isotopic exchange fractionation depending on temperature and cooling time. However, none of the bulk samples analysed showed any systematic variation in $\delta^{34}S$, indicating no loss or gain of sulphur due to changes in mineralogy.

The $\delta^{34}S(S^=)$ distribution patterns obtained for Isua sediments, and intercalated amphibolites broken down into the various facies 'oxide', 'silicate', 'siderite', 'sulphide', are shown in Figure 3. The most striking features displayed by the $\delta^{34}S$ values of the Isua

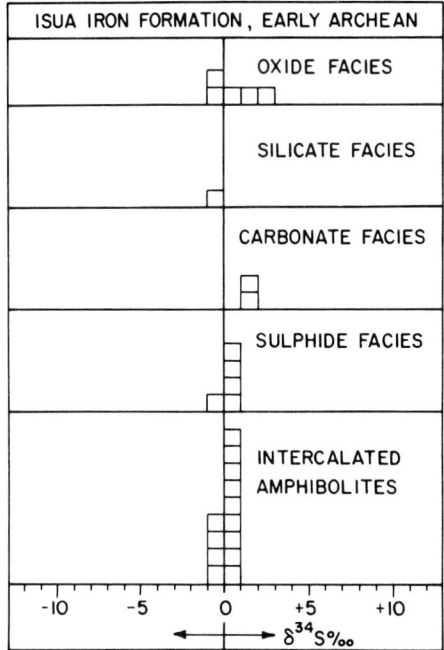

Fig. 3. $\delta^{34}S\%_{oo}$ C.D.T. distribution patterns of sulphides in samples collected from the various iron facies, (oxide, silicate, carbonate, sulphide) and intercalated tuffaceous amphibolites of the Isua banded iron-formation sediments in West Greenland. Early Archean 3.7×10^9 yr (Moorbath et al., 1973; Monster et al., 1979).

sediments are the small variations in $\delta^{34}S$, ($\pm 0.5\%_{oo}$) and the small displacements of mean values from zero ($+0.5\%_{oo}$). In this regard the intercalated tuffaceous amphibolites show the same displacement from zero as the banded iron-formation rocks. The very small variations and displacements in $\delta^{34}S$ that are observed can easily be accounted for by temperature and differentiation isotope effects. These results are in marked contrast to the sulphur isotope distribution patterns for the 2.75×10^9 yr old Michipicoten and Woman River iron-formations of the Canadian Shield which have yielded a fairly widespread (and variable) $\delta^{34}S$ distribution pattern consistent with isotope fractionation processes in the bacterial reduction of sulphate.

The absence of isotope distribution patterns characteristic of bacterial reduction processes in an ancient sedimentary rock does not preclude the presence of such patterns in other sedimentary rocks of the same age. However, it seems unlikely that evidence will be found in other sediments of Isua age in view of the extent of the Isua banded iron-formation sediments and the variety of environmental conditions under which the various facies were deposited (Monster et al., 1979).

Monster et al. (1979) also considered the possibility of dissimilatory bacterial reduction of sulphate taking place under extreme conditions prevailing in Archean time without giving rise to isotope fractionation.

In laboratory experiments the isotope effect does decrease toward zero at very high metabolic rates under optimum conditions and in the very low sulphate concentration range $(SO_4^=) \cong 10^{-4}$ to 10^{-5} molar (Harrison and Thode, 1958). However, these extreme situations are probably not relevant to natural situations.

In conclusion, there is no isotopic evidence that sulphur of the Isua iron-formation has passed as a result of dissimilatory sulphate reduction through to H_2S with subsequent precipitation of sedimentary sulphide.

The complete lack of evidence for dissimilatory 'sulphate reducers' in the Isua banded iron-formation early Archean, and the overwhelming evidence both from isotopic and chemical considerations for their presence in banded iron-formations of the Canadian Shield late Archean, places early and late time limits on the time of their beginnings.

References

Appel, P. W. U.: 1979, *Econ. Geol.* **74**, 45.
Burnie, S. W., Schwarcz, H. P., and Crocket, J. H.: 1972, *Econ Geol.* **67**, 895.
Cnukhrov, F. V., Vinogradov, V. I., and Ermilova, L. P.: 1970, *Mineralium Deposita* **5**, 209.
Garrels, R. M., Perry, E. A. Jr., and MacKenzie, F. T.: 1973, *Econ. Geol.* **68**, 1173.
Goldich, S. S.: 1973, *Econ. Geol.* **68**, 1126.
Goodwin, A. M.: 1964, *Econ. Geol.* **59**, 684.
Goodwin, A. M.: 1973, *Econ. Geol.* **68**, 915.
Goodwin, A. M., Monster, J. and Thode, H. G.: 1976, *Econ. Geol.* **71**, 870.
Harrison, A. G. and Thode, H. G.: 1958, *Trans. Faraday Soc. No. 421*, **54**, 84.
Heinrichs, T. K. and Reimer, T. O.: 1977, *Econ. Geol.* **72**, 1426.
Holland, H. D.: 1973, *Econ. Geol.* **68**, 1169.
Holser, W. T. and Kaplan, I. R.: 1966, *Chem. Geol.* **1**, 93.
Ishii, M. M.: 1953, Master's Thesis, McMaster University, Hamilton. Ontario.
James, H. L.: 1954, *Econ. Geol.* **49**, 235.
Kaplan, I. R. and Rittenberg, S. C.: 1964, *J. Gen. Microbiol.* **34**, 195.
Kaplan, I. R., Emery, K. O., and Rittenberg, S. C.: 1963, *Geochim. Cosmochim. Acta.* **27**, 297.
Kemp, A. L. W. and Thode, H. G.: 1968, *Geochim. Cosmochim. Acta* **32**, 71.
Monster, J. and Thode, H. G.: 1978, (unpublished).
Monster, J., Appel, P. W. U., Thode, H. G., Schidlowski, M., Carmichael, C. M., and Bridgwater, D.: 1979, *Geochim. Cosmochim. Acta* **43**, 405.
Moorbath, S., O'nions, R. K., and Parkhurst, R. T.: 1973, *Nature* **245**, 138.
Perry, E. C. Jr., Monster, J., and Reimer, T.: 1971, *Science* **171**, 1015.
Rees, C. E.: 1969, *Earth Planet. Sci. Letters* **7**, 366.
Shima, M., Gross, W. H., and Thode, H. G.: 1963, *J. Geophys. Res. No. 9*, **68**, 2835.
Thode, H. G.: 1970, *Mineral. Soc. Am. Spec. Paper* **3**, 133.
Thode, H. G. and Monster, J.: 1965, *Symp. Memoir No. 4, A.A.P.G.*, 367.
Thode, H. G. and Monster, J.: 1979, (unpublished).
Thode, H. G., Kleerekoper, H., and McElcheran, D. E.: 1951, *Research* **4**, 581.
Thode, H. G., Harrison, A. G., and Monster, J.: 1960, *Bull. A.A.P.G.* **44**, 1809.
Thode, H. G., Monster, J., and Dunford, H. B.: 1961, *Geochim. Cosmochim. Acta* **25**, 159.
Thode, H. G., Dunford, H. B., and Shima, M.: 1962, *Econ. Geol.* **57**, 565.
Vinogradov, A. P., Grinenko, V. A., and Ustinov, V. I.: 1962, *Geokhimiza* **10**, 973.
Vinogradov, V. I., Reimer, T. O., Leites, A. M., and Smelov, S. B.: 1976, *Litol. Polezn Iskop* **11**, 12.
Wanless, R. K., Boyle, R. W., and Lowdon, J. A.: 1960, *Econ. Geol.* **55**, 1591.

ANTIQUITY AND EVOLUTIONARY STATUS OF BACTERIAL SULFATE REDUCTION: SULFUR ISOTOPE EVIDENCE

MANFRED SCHIDLOWSKI

Max-Planck-Institut für Chemie (Otto-Hahn-Institut) Saarstr. 23, D-6500 Mainz, W-Germany*

Abstract. The presently available sedimentary sulfur isotope record for the Precambrian seems to allow the following conclusions: (1) In the Early Archaean, sedimentary $\delta^{34}S$ patterns attributable to bacteriogenic sulfate reduction are generally absent. In particular, the $\delta^{34}S$ spread observed in the Isua banded iron formation (3.7×10^9 yr) is extremely narrow and coincides completely with the respective spreads yielded by contemporaneous rocks of assumed mantle derivation. Incipient minor differentiation of the isotope patterns notably of Archaean sulfates may be accounted for by photosynthetic sulfur bacteria rather than by sulfate reducers. (2) Isotopic evidence of dissimilatory sulfate reduction is first observed in the upper Archaean of the Aldan Shield, Siberia ($\sim 3.0 \times 10^9$ yr) and in the Michipicoten and Woman River banded iron formations of Canada (2.75×10^9 yr). This narrows down the possible time of appearance of sulfate respirers to the interval $2.8 - 3.1 \times 10^9$ yr. (3) Various lines of evidence indicate that photosynthesis is older than sulfate respiration, the SO_4^{2-} utilized by the first sulfate reducers deriving most probably from oxidation of reduced sulfur compounds by photosynthetic sulfur bacteria. Sulfate respiration must, in turn, have antedated oxygen respiration as O_2-respiring multicellular eucaryotes appear late in the Precambrian. (4) With the bulk of sulfate in the Archaean oceans probably produced by photosynthetic sulfur bacteria, the accumulation of SO_4^{2-} in the ancient seas must have preceded the buildup of appreciable steady state levels of free oxygen. Hence, the occurrence of sulfate evaporites in Archaean sediments does not necessarily provide testimony of oxidation weathering on the ancient continents and, consequently, of the existence of an atmospheric oxygen reservoir.

1. Introduction

The search for the antiquity, and the assessment of the evolutionary status, of dissimilatory sulfate reduction ('sulfate respiration') are important topics in current efforts to elucidate the early history of life (cf. Trudinger, 1976). There is no doubt that the advent of sulfate respiration must have constituted a paramount quantum step in bioenergetic evolution and, accordingly, a most important benchmark in the temporal framework of early organic evolution as a whole. Although the sequential relationship of the principal steps in the evolution of the respiratory pathway is still under debate (Egami, 1974, 1976; Broda, 1975a, 1977), we may reasonably assume that, as a result of photosynthetic activity by green and purple sulfur bacteria, sulfate as a mild oxidant must have preceded free oxygen in the ancient environment (Broda, 1975b, p. 77). Hence, SO_4^{2-} as a terminal electron acceptor for the oxidation of organic substrates ('foodstuffs') should have become available in the oceans long before the attainment of appreciable steady state

---

* Current address: Precambrian Paleobiology Research Group, Dept. of Earth and Space Sciences, UCLA, Los Angeles, Calif. 90024, U.S.A.

levels of molecular oxygen in the atmosphere, with sulfate respiration thus certainly antedating the process of O_2-respiration.

In contrast to nitrate respirers (whose evolutionary status dominates current bioenergetic discussions), sulfate reducing bacteria can be traced back in the geological record as they release H_2S as a principal metabolite which is susceptible to subsequent fixation as sedimentary sulfide (mostly as pyrite, FeS_2). Moreover, during bacterial sulfate reduction, i.e.,

$$2 CH_2O + SO_4^{2-} \rightarrow 2 CO_2 + H_2S + 2 OH^- \qquad (1)$$

the stable isotopes of sulfur are selectively metabolized, with ^{32}S preferentially accumulating in bacteriogenic H_2S as a result of a kinetic isotope effect inherent in this process. In terms of the conventional δ-notation, the $\delta^{34}S$ values of metabolic H_2S are markedly shifted to the negative side when compared to those of the parent sulfate pool, the magnitude of the effect ranging from a few per mil. to some 50‰ as a result of its dependence on a complex set of variables which govern the rate of reduction per bacterial cell (see, inter alia, Kaplan and Rittenberg, 1964; Kemp and Thode, 1968). Accordingly, the prime characteristics of the dissimilatory reduction of sulfate by sulfate reducers are (1) a general enrichment of the light sulfur isotope (^{32}S) and (2) a relatively large spread in $\delta^{34}S$ values of the resulting metabolic H_2S-phase.

Both these characteristics are basically preserved when hydrogen sulfide is incorporated in sulfide minerals. This happens in many sedimentary environments, notably in the 'euxinic' facies associated with sulfuretum-type anaerobic ecosystems. Thus, sedimentary sulfides inherit the characteristic isotope spread of bacteriogenic H_2S, providing mineralized vestiges of former life activities of microbial sulfate reducers. It has been shown, furthermore, that primary $\delta^{34}S$ gradients are unlikely to be smoothed out by subsequent metamorphic processes, even in high-grade terranes (Buddington et al., 1969; Rye and Ohmoto, 1974).

During the last decade or so, sulfur isotope have accrued for a fair amount of Precambrian and notably Archaean sediments which seem to suggest certain limits for the time of emergence of dissimilatory sulfate reduction (see, inter alia, Thode et al., 1962; Perry et al., 1971; Vinogradov et al., 1976; Goodwin et al., 1976; Donnelly et al., 1977). The recent accumulation of the first sulfur isotope values (Monster et al.,

---

Fig. 1. Isotopic composition of sulfides from igneous rocks of Archaean age as compared with sulfur from the troilite phase of meteorites. The narrow clustering of $\delta^{34}S$ values around zero per mil. as observed in several mafic bodies [(2), (3)] indicates 'primitive' sulfur of mantle derivation, while the moderate spreads displayed by (4), (5) and (6) reflect already incipient isotope differentiation (either by intramagmatic fractionation processes or as a result of contamination with sedimentary sulfur). – (1) Troilite in various types of meteorites (Hulston and Thode, 1965); (2) sulfides in Isua orthoamphibolites (3.76 x 10^9 yr) of tholeiitic affinities (Monster et al., 1979); (3) sulphides in Ameralik basalts (between 3.1 and 3.7 x 10^9 yr) of Isua region (Monster et al., 1979); (4) sulfides in mafic and ultramafic intrusions at base of Archaean Iengra Series (~3.5 x 10^9 yr), Aldan Shield, Siberia (Vinogradov et al., 1976); (5) sulfides in felsic volcanics intercalated in Michipicoten banded iron formation of Canada (Goodwin et al., 1976); (6) nickel sulfides in ultramafic/mafic sequences (2.6–2.8 x 10^9 yr) of Yilgarn Block, Australia (Donnelly et al., 1977).

1979) from the oldest terrestrial sediments from Isua, West Greenland (3.76 × 10^9 yr) (Moorbath *et al.*, 1973) as well as from other Archaean sequences (Fripp *et al.*, 1978; Lambert *et al.*, 1979) makes this perhaps an appropriate time for a review of the ancient sulfur cycle. In the following, an overview will be given on the presently known

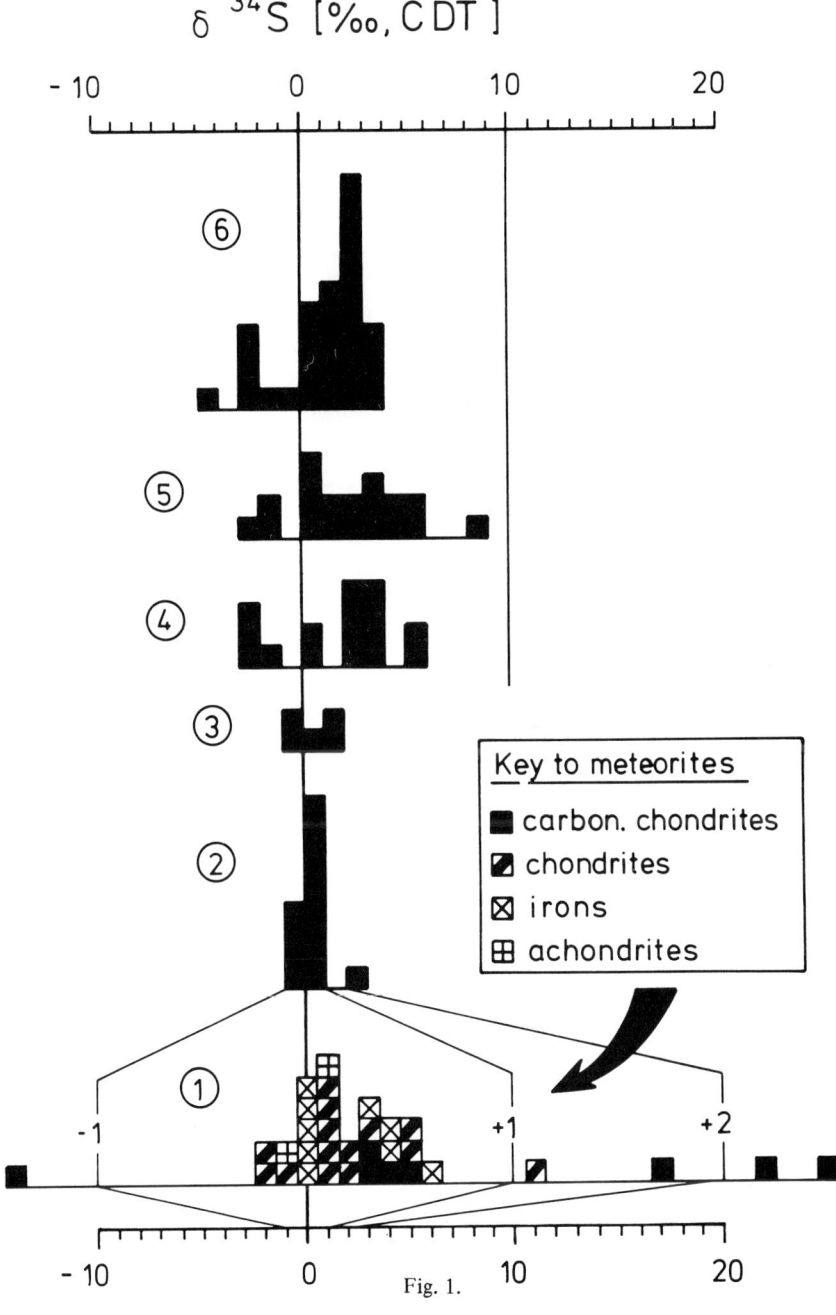

Fig. 1.

Precambrian sulfur isotope record which — though incomplete over large periods — should allow us to decidedly narrow in time the possible rise of sulfate respirers during the Early Precambrian.

2. Isotopic Composition of Sulfur in Archaean Primary Rocks

The sulfur isotope abundances observed in several mafic igneous rocks of Archaean age (see Figure 1) provide a natural base-line for the interpretation of the early sedimentary sulfur isotope record. Sulphur with $\delta^{34}S$ values close to zero per mil. as encountered in sulfides from both the 3.7×10^9 yr-old Isua amphibolites of tholeiitic affinities and the somewhat younger (between 3.1 and 3.7×10^9 yr) Ameralik basalts of the same region [see Figure 1, (2), (3)] is generally supposed to reflect the average isotopic composition of mantle sulfur, coming very close to sulfur in the troilite phase meteorites. It is reasonable to assume that isotopic differentiation of terrestrial sulfur had originally started with primordial sulfur of this type. As is attested by the sizeable spread around zero per mil. of the $\delta^{34}S$ values yielded by several primary rocks [see Figure 1, (4), (5), (6)], isotopic fractionation of primitive sulfur may actually commence at a very early stage, either as a result of intramagmatic differentiation processes or by assimilation of isotopically fractionated sulfur from the surrounding sediments.

3. Precambrian Sedimentary Sulfur Isotope Record

Figure 2 is a compilation of the principal sedimentary sulfur isotope data (for sulfide and sulfate) hitherto available for the time span between the beginning of the record some 3.7×10^9 yr ago and Grenville age (about 10^9 yr) Except for those from the Grenville Province, these data reflect the isotopic geochemistry of sulfur from *common* sedi-

---

Fig. 2. Sedimentary sulfur isotope record for the time range $3.7-1.0 \times 10^9$ yr. ago (time axis not to scale; note that (2), (3) and (4) are approximately coeval). — (1) Isua, West Greenland, banded iron formation (3.76×10^9 yr); (2) Pilbara Block, Australia ($\sim 3.5 \times 10^9$ yr); sulfate occurring as bedded barite); (3) Fig Tree Group, Swaziland System, South Africa ($3.0-3.3 \times 10^9$ yr; sulfate occurring as bedded barite); (4) Archaean of Aldan Shield, Siberia (lower suites approaching 3.5×10^9 yr): (a) Upper Aldan and Nimnyr Suites of Iengra Series, (b) and (c) Fedorov Suite of Iengra Series (sulfate occurring as barite and anhydrite), (d) and (e) Taeshnoe and Pionerskoe iron ore deposits in Fedorov Suite of Iengra Series, (f) and (g) Dsheltula Series; (5) various banded iron formations from Rhodesia (mostly $> 2.8 \times 10^9$ yr); (6) black shales from greenstone belts within Yilgarn Block, Australia ($2.6-2.8 \times 10^9$ yr); (7) and (8) Michipicoten and Woman River banded iron formations, Superior Province, Canada ($\sim 2.75 \times 10^9$ yr); (9) black shales from Outokumpu, Finland (between 1.8 and 2.3×10^9 yr), (10) Sudbury Precambrian: (a) Frood Series ($\sim 2.2 \times 10^9$ yr), (b) Onwatin Slates of Sudbury basin ($\sim 1.9 \times 10^9$ yr); (11) Grenville Province of Canadian Shield ($1.0-1.2 \times 10^9$ yr): (a) Adirondack sedimentary sulfide deposits, (b) sphalerite-pyrite deposits of Balmat-Edwards, (c) bedded anhydrite, Balmat-Edwards. — Data from Brown, 1973 (11b,c); Buddington *et al.*, 1969 (11a); Donnelly *et al.*, 1977 (6); Fripp *et al.*, 1978 (5); Goodwin *et al.*, 1976 (7,8); Lambert *et al.*, 1978 (2); Mäkelä, 1974 (9); Monster *et al.*, 1979 (1); Perry *et al.*, 1971, 1975 (3,2); Thode *et al.*, 1962 (10); Vinogradov *et al.*, 1976 (3,4).

mentary environments, with major stratiform ore bodies (e.g., Broken Hill, Mt. Isa) and assemblages of detrital sulfides (Witwatersrand) deliberately excluded (for a review of the isotope geochemistry of these latter occurrences see Schidlowski, 1973).

As is obvious from Figure 2, the isotope data from the Isua banded iron formation

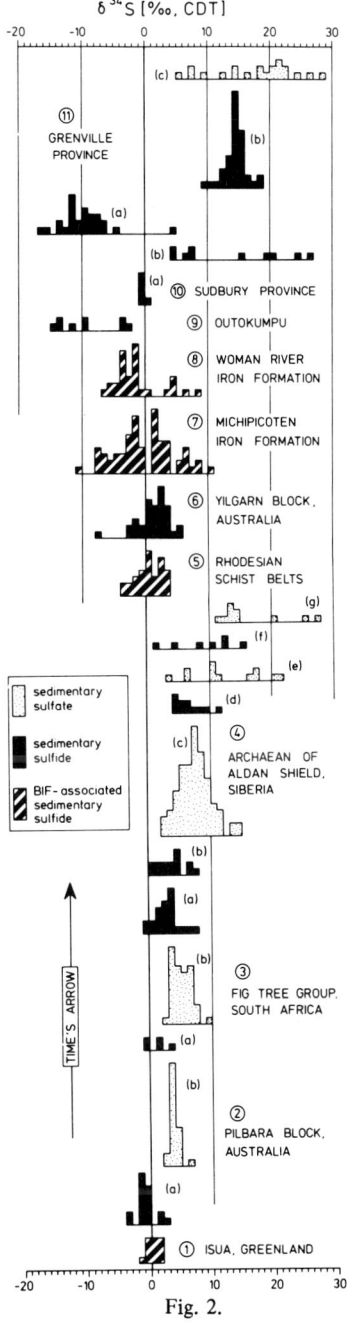

Fig. 2.

represent the lowermost benchmark in the sedimentary sulfur isotope record. With an average $\delta^{34}S$ of +0.5 ± 0.9‰ [CDT] (Monster et al., 1979), the sulfides of the various BIF-members investigated (belonging mostly to the sulfide and oxide facies) are virtually identical in their isotopic composition with magmatogenic sulfur as present, for instance, in the tuffaceous Isua amphibolites ($\delta^{34}S$ = +0.3 ± 0.9‰) and the basaltic Ameralik dykes of the Isua region [$\delta^{34}S$ = +0.6 ± 1.1‰; cf. Figure 1, (2), (3)].| Hence, the sulfide constituents of the Isua banded iron formation are clearly made up of very primitive sulfur resembling primordial sulfur from the Earth's mantle (cf. Shima et al., 1963; Schneider, 1970). The most probable source of this sulfur were volcanogenic sulfide emanations which, during the passage to their final sites of emplacement within the sedimentary sequence, have experienced minor isotopic fractionation accounting for the observed maximum spread of $\delta^{34}S$ values between −1.0 and +2.0‰. It should be noted that there is no isotopic evidence whatsoever that the sulfur of the Isua iron formation has passed through a stage of bacteriogenic H_2S with subsequent precipitation as sedimentary sulfide. Accordingly, the observed sulfur isotope distribution in these sulfides would be consistent with the absence of microbial sulfate respirers prior to some 3.7×10^9 yr ago.

Such a conclusion is apparently substantiated by a triad of somewhat younger occurrences of sedimentary sulfur which all fall roughly into the time interval 3.0 to 3.5×10^9 yr [Pilbara Block, Australia; Barberton greenstone belt, South Africa; Aldan Shield, Siberia; see Figure 1, (2), (3) and (4)]. These occurrences are characterized by the coexistence in the same sedimentary sequence of both sulfide and sulfate, the latter occurring mostly as barite currently interpreted as either an original relic or a later replacement of the oldest evaporite series preserved in the sedimentary column (Perry et al., 1971; Vinogradov et al., 1976; Lambert et al., 1979). Although the sedimentary origin notably of the Barberton barites has been repeatedly questioned, there seems to be convincing evidence by now that these sulfates were originally deposited as chemical sediments (Heinrichs and Reimer, 1977). Further, Lowe and Knauth (1977) have reported from the underlying Onverwacht cherts abundant spindle-shaped crystallites resembling pseudomorphs after gypsum ('gypsum ghosts'). From the barite-chert series of the North Pole (Pilbara) deposits Dunlop (1978) has also described textural features suggesting that the barites were essentially formed by diagenetic replacement of an evaporitic calcium sulfate precursor by Ba^{2+}-bearing intrastratal solutions (without significantly affecting the bulk isotope geochemistry of the sulfates). The original precipitation of a fairly soluble calcium sulfate phase would, necessarily, imply relatively high SO_4^{2-}-concentrations in the Archaean oceans, this lending further support to concepts of a conservative sea water chemistry through the ages (cf. Garrels and Mackenzie, 1974; Holland, 1974).

A conspicuous feature of these Early Archaean sulfur occurrences is the very small isotopic fractionation between sulfide and sulfate amounting to only 3 to 4‰ on average. This contrasts markedly with fractionations of 30 ± 5‰ encountered in younger, and notably Phanerozoic, sediments (cf. Holser and Kaplan, 1966; Holland, 1973;

Schidlowski et al., 1977). With their $\delta^{34}S$ means lying between +4 and +7‰, the sulfates are not too far in their isotopic composition from primordial sulfur, while the sulfides largely fall into the range of this latter. It would seem reasonable, therefore, to concur in the statement by Lambert (1978) that juvenile sulfur with $\delta^{34}S \cong 0\%$ was the most probable source of *both* sulfur species.

If sulfide and sulfate in these sediments are *genetically related* (which interpretation would exclude, for instance, the presence of substantial amounts of detrital pyrite), then the sulfur isotope geochemistry of these Archaean sequences could, in principle, be explained in two alternative ways:

1. First, the minor shift of the $\delta^{34}S$ values from zero to averages between +3 and +7‰ as displayed by the Early Archaean sulfates might be interpreted as heralding the beginnings of bacterial sulfate reduction (which, when in full control of the exogenous sulfur cycle, will push the mean for sulfate evaporites much further into the positive field). However, the isotopic composition of sulfides in the enclosing sediments is not consistent with this explanation, showing neither the characteristic magnitude of fractionation between bacteriogenic sulfide and the contemporaneous sulfate source nor the distribution pattern typical of bacteriogenic $\delta^{34}S$ values. Although experimental work by Harrison and Thode (1957) has shown that, under specific conditions (notably low sulfate concentrations of the order of 10^{-5} molar), the isotope effect in bacterial sulfate reduction may indeed approach zero, it seems unlikely that such conditions are relevant to our case (with evidence piling up in favor of an original presence in these series of calcium sulfate which would necessitate fairly high SO_4^{2-}-concentrations in the waters of the evaporitic basin). Altogether, the $\delta^{34}S$ distribution patterns displayed by the sedimentary sulfides certainly have closer affinities to magmatic sulfur than to the respective patterns of bacteriogenic sulfides.

2. An alternative explanation may be based on the assumption that the bulk of the sulfate ions stems from photosynthetic oxidation of primordial (reduced) sulfur by green and purple sulfur bacteria. With $SO_2 \ll H_2S$ in juvenile sulfur emanations, and SO_2 just ending up as sulfite (SO_3^{2-}) in low redox environments, the formation of sulfate will be contingent upon a subsequent oxidation of reduced sulfur compounds as well as of SO_2 to the sulfate stage. Although local inorganic oxidation processes cannot be completely discounted even under the reduced conditions prevalent on the primitive Earth, the most important contribution towards the early terrestrial SO_4^{2-}-budget was probably made by photosynthetic sulfur bacteria (Broda, 1975b, p. 77). Providing optimum conditions for the proliferation of anaerobic microbial ecosystems, the Archaean might have even been the 'golden age' of green and purple sulfur bacteria. In processes of bacterial photosynthesis, e.g.,

$$\tfrac{1}{2}H_2S + CO_2 + H_2O \xrightarrow{h\nu} CH_2O + H^+ + \tfrac{1}{2}SO_4^{2-} \qquad (2)$$

minor to moderate enrichments of the heavy sulfur isotope have been observed in the resulting sulfate and polythionate fractions, the effect being usually very small (0–1‰) for SO_4^{2-}, but significant (up to about 11‰) for the transient (metastable) $S_xO_6^{2-}$-phase

(Kaplan and Rittenberg, 1964). As already proposed by Perry et al. (1971), the moderately positive $\delta^{34}S$ values displayed by the bulk of the Archaean barite deposits could, therefore, be consistent with an origin of their SO_4^{2-}-component by photosynthetic oxidation of reduced primordial sulfur.

In summary, the isotopic geochemistry of these Archaean sulfide-sulfate pairs seems to lend reasonable support to the assumption that photosynthetic sulfur bacteria were already thriving in the ancient seas, whereas there is little (if any) unequivocal evidence for the presence of sulfate respirers. On the other hand, incipient differentiation of $\delta^{34}S$ distribution patterns for both sulfide and sulfate has been reported for the upper Archaean suites of the Aldan Shield (Vinogradov et al., 1976), the respective histograms (see Figure 2, 4d–g) showing indeed some salient features of bacteriogenic patterns. Since geological field evidence indicates, however, extensive metamorphic and metasomatic reconstitution of the host rocks of these sulfide-sulfate pairs, it is proposed that not too much significance be attached to these data at the present stage.

Any large-scale bacterial sulfate reduction is also discounted by the $\delta^{34}S$ distribution ($\delta^{34}S$ = +0.6 ± 2.0‰) displayed by sulphides from various banded iron formations from the Archaean schist belts of Rhodesia [Fripp et al., 1978; see Figure 2, (5)]. As in the case of the Isua iron formation, there is no doubt that these sulfides are made up of 'primitive' sulfur of magmatic pedigree. The bulk of the Rhodesian samples stem from the Early Archaean Sebakwian Group underlying the Bulawayan, the latter having been recently dated at 2.6–2.8 x 10^9 yr by Hawkesworth et al. (1975).

A plot of 35 sulfide values from black shales within the 2.6–2.8 x 10^9 yr old greenstone belts of the Australian Yilgarn Block (Donnelly et al., 1977) also shows a marked preponderance of magmatogenic sulfur [cf. Figure 2, (6)]. Although occurring as a constituent of typically 'euxinic' sediments, there is little isotopic evidence that the bulk of these sulfides has ever passed through a stage of bacteriogenic H_2S with subsequent precipitation as sedimentary sulfide. We would, therefore, concur in the interpretation by Donnelly et al. (op. cit.) that hydrothermal emanations must have constituted the principal sulfur source of these sulfides. It cannot be excluded, however, that some rare negative $\delta^{34}S$ values encountered possibly indicate minor contributions by biological sulfate reduction.

The decisive break in the presently known sedimentary sulfur isotope record is indicated by the sulfide patterns yielded by the 2.75 x 10^9 yr old Michipicoten and Woman River banded iron formations of Canada (Goodwin et al., 1976). Here, both the large spread of $\delta^{34}S$ values over some 20‰ and their marked encroachment upon the negative field [Figure 2, (7), (8)] leave hardly any doubt that these are genuinely bacteriogenic patterns (perhaps with some superimposed magmatogenic values responsible for the peaks in the zero permil range). Variations of such patterns are shown by most younger Precambrian sedimentary sulfides hitherto investigated, notably those from black shales occurring in the immediate vicinity of the Outokumpu deposit, Finland (between 1.8 and 2.3 x 10^9 yr, cf. Mäkelä, 1974) and from the 1.0–1.2 x 10^9 yr old Grenville Province of the Canadian Shield [Figure 2, (9), (11)]. It is worth noting that, during

Grenville times, isotope values of seawater sulfate were rather modern as is attested by a $\delta^{34}S$ average of about 21‰ [Figure 2, (11c)] yielded by the bedded anhydrite in the surroundings of the Balmat-Edwards sphalerite-pyrite deposit (Brown, 1973). The similarity of the isotope geochemistry of the Balmat-Edwards sulfides [cf. Figure 2, (11b)] with the sulfates of their sedimentary frame is generally explained in terms of a total conversion into sulfide, by bacterial sulfate reduction, of a limited sulfate reservoir of above composition. In contrast, the Adirondack sulfide deposits of the same province (Buddington et al., 1969) display a bacteriogenic isotope pattern typical of sulfate reduction in open systems, combining a relatively large spread of $\delta^{34}S$ values with an average lying well in the negative field [Figure 2, (11a)].

The two sulfide occurrences from the Sudbury district (Thode et al., 1962) included in Figure 2 (10a and 10b) may serve as an example of sedimentary sulfur of different pedigree occurring in relatively close spatial and temporal vicinity. While the sulfides from the Frood Series (10a) south of the Sudbury basin (with $\delta^{34}S$ values close to zero per mil.) are obviously made up of primary sulfur derived from local volcanogenic sources, those from the Onwatin slates (10b) inside the basin display the wide isotope spread typical of biogenic material.

4. Time of Emergence and Evolutionary Status of Dissimilatory Sulfate Reduction

With the presently available Precambrian sulfur isotope record at hand (Figure 2), an attempt to determine the possible time of emergence of bacterial sulfate reduction seems to be called for. Because of the limits imposed by the scanty data base and the poor time resolution (notably for the Archaean part of the record), this event can be narrowed down at present only to a time interval of about 300–400 million years.

As already proposed elsewhere (Lambert, 1978; Monster et al., 1979), the isotope distribution patterns of the sulfides from the Michipicoten and Woman River iron formations of Canada (2.75×10^9 yr) are likely to provide the oldest clear-cut evidence of bacterial sulfate reduction. On the other hand, the triad of Early Archaean sulfide-sulfate pairs represented in Figure 2 (2, 3, 4b and c) does not furnish convincing proof of contemporaneous sulfate respirers as has been fully set out before. Accordingly, the rise of microbial sulfate reducers should have antedated the Michipicoten and Woman River iron formations and postdated the sulfate occurrences of the Pilbara Block, the Barberton greenstone belt and the Iengra Series of the Aldan Shield respectively, thus leaving the time interval between 2.8 and some 3.1×10^9 yr as the most probable age limit for this event. Consequently, efforts to further narrow down in time this important step in bioenergetic evolution should concentrate on this particular time span.

The absence of bacteriogenic isotope patterns in the Isua iron formation decidely fits into the general trend of the Archaean $\delta^{34}S$ record. Although the lack of bacteriogenic features in one particular sedimentary environment does, per se, not preclude bacterial activity in coeval sediments, it would be unrealistic, in view of the above findings, to expect evidence of biological sulfur isotope fractionation as early as 3.7×10^9 yr ago. Because of this consistency with the Archaean sulfur isotope record as a whole, the interpretation of the Isua data in terms of an absence of bacterial sulfate reduction can be hardly invalidated by the occasional encounter of 'primitive' isotope patterns in geologically younger sediments and notably BIF series [caused by a preponderance in local environments of hypogenic sulfur; see, for instance, Figure 2, (5) and (10a)].

Hence, sedimentary sulfur isotope data presently available for the Archaean would suggest that sulfate respiration was a relatively late achievement in the evolution of bioenergetic processes. In particular, there is hardly any doubt that photosynthesis must have preceeded sulfate respiration. This is prompted by both paleontological (Muir and Grant, 1976; Knoll and Barghoorn, 1977) and goechemical evidence (Junge *et al.*, 1975; Sidorenko and Sidorenko, 1975), with latest investigations dating back the probable beginnings of photosynthesis to at least Isua times (Schidlowski *et al.*, 1979). It is in keeping with such conjectures that photosynthetic (S-oxidizing) sulfur bacteria are presently being regarded as the most likely ancestors of sulfate reducers (Peck, 1974; Broda, 1975b, p. 112). This notion is supported, inter alia, by sequence analysis of ferredoxins from photosynthetic and sulfate-reducing sulfur bacteria (Schwartz and Dayhoff, 1978). In the phylogenetic tree of ferredoxins, photosynthetic sulfur bacteria diverge from primitive anaerobic heterotrophys (like *Clostridium*) at a very early stage, while sulfate reducers later branch from the base of the line leading to the blue-green algae (Figure 3).

Since the availability of sufficient quantities of sulfate is a necessary prerequisite for dissimilatory sulfate reduction, we may infer that photosynthetic sulfur bacteria must have set the stage for sulfate reducers, the advent of the latter heralding the establishment of a complete biological sulfur cycle ultimately powered by solar radiation. Since O_2-respiring multicellular eucaryotes have appeared late in the Precambrian (cf. Cloud, 1976; Schopf, 1974), there is no doubt that sulfate respiration has antedated oxygen respiration, the most important quantum step in bioenergetic evolution (Schidlowski, 1978). The assumed sequence sulfate respiration → oxygen respiration would necessarily imply that sulfate had become available in the ancient environment considerably earlier than free oxygen which would follow directly from the priority of bacterial photosynthesis to the more differentiated (O_2-releasing) cyanophytic process.

5. Environmental Significance of Archaean Sulfate Evaporites

Since the bulk of sulfate participating in the *present* geochemical cycle unquestionably owes its origin to oxidation weathering of reduced sulfur compounds, the occurrence of bedded sulfates in Archaean sediments has been repeatedly taken as testimony of the existence of a contemporaneous atmospheric oxygen reservoir (Chukhrov *et al.*, 1970; Vinogradov *et al.*, 1976). However, with photosynthetic sulfur bacteria extant during the Archaean, inferences along these lines should be treated with due reserve.

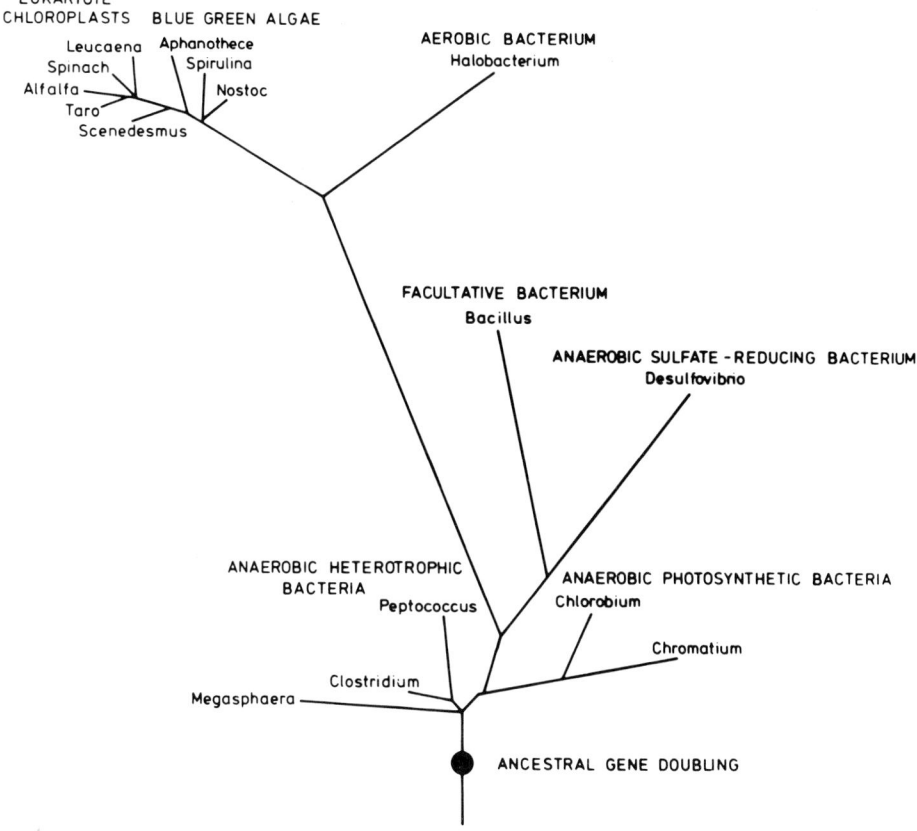

Fig. 3. Simplified evolutionary tree of ferredoxins (adapted from Schwartz and Dayhoff, 1978). Note that photosynthetic sulfur bacteria (*Chlorobium*, *Chromatium*) have branched very early from heterotrophic anaerobes (*Megasphaera*, *Clostridium*, *Peptococcus*) whose ferredoxins show the strongest evidence of the ancestral gene doubling observed in all ferredoxin sequences (hence, this event must have antedated the species divergences pictured in the tree). The *Desulfovibrio* line, on the other hand, has diverged from the root of the subtree which has consequently branched into aerobic bacteria, blue-green algae (cyanophytes), and plants.

As already pointed out above (Equation 2), photosynthetic bacteria utilize hydrogen sulfide (primarily H_2S from volcanic sources) instead of water as electron donor for the reduction of carbon dioxide. Since hydrogen sulfide was probably more abundant in the reducing environment of the Early Archaean than it is today, it may not be unreasonable to look at this time as the possible 'golden age' of photosynthetic sulfur bacteria. With the bulk of sulfate within the Archaean oceans stemming almost certainly from the life activities of these photolithotrophs, SO_4^{2-} as a mild oxidant should have become plentiful in the ancient seas long before the attainment of appreciable steady-state levels of molecular oxygen in the atmosphere (cf. Broda, 1975b, p. 77). Consequently, the occurrence of sulfate evaporites in Archaean sediments does not necessarily constitute evidence of oxidation weathering on the ancient continents and thus of the existence of a contemporaneous atmospheric oxygen reservoir.

On the other hand, the isotopic composition of the oldest sedimentary carbonates (Schidlowski *et al.*, 1979) suggests a rather modern ratio of organic carbon to carbonate carbon in the Earth's sedimentary shell already some 3.7×10^9 yr ago, this implying the existence of a sizeable reservoir of oxidation equivalents of the reduced carbon constituents. If we are correct in assuming that the presence in the Isua supracrustals of oxide-facies banded iron formation is indicative of cyanophytic (O_2-releasing) photosynthesis, then a major part of these oxidation equivalents could have been molecular oxygen. Accordingly, free oxygen could have indeed appeared very early in the Earth's history, but is unlikely to have accumulated in the atmosphere before the saturation of the principal oxygen sinks (notably the Fe^{2+}-burden of the ancient seas). An inconsistency between the position of cyanophytes in the evolutionary scheme of Figure 3 and their inferred presence during Isua times would deserve further inquiry.

Acknowledgments

This work was performed as part of the Sonderforschungsbereich 73 ('Atmospheric Trace Components'), receiving partial funding from the Deutsche Forschungsgemeinschaft. It has substantially benefited from discussions with, or comments by, Margaret Dayhoff, C. E. Junge, H. G. Thode and P. A. Trudinger. T. H. Donnelly was so kind as to make available to me the Pilbara data of Lambert *et al.* (1978) prior to publication.

References

Broda, E.: 1975a, *J. Mol. Evol.* 7, 87.
Broda, E.: 1975b, *The Evolution of the Bioenergetic Processes*, Pergamon, 220 pp.
Broda, E.: 1977, *Origins of Life* 8, 173.
Brown, J. S.: 1973, *Econ. Geol.* 68, 362.
Buddington, A. F., Jensen, M. L. and Mauger, R. L.: 1969, *Geol. Soc. Amer. Mem.* 115, 423.
Chukhrov, F. V., Vinogradov, V. I. and Ermilova, L. P.: 1970, *Mineral. Deposita* 5, 209.
Cloud, P. E.: 1976, *Paleobiology* 2, 351.
Donnelly, T. H., Lambert, I. B., Oehler, D. Z., Hallberg, J. A., Hudson, D. R., Smith, J. W., Bavinton, O. A. and Golding, L.: 1977, *J. Geol. Soc. Austr.* 24, 409.
Dunlop, J. S. R.: 1978, *Publ. Geol. Dept. & Extension Service, Univ. West. Austr.* 2, 30.
Egami, F.: 1974, *Origins of Life* 5, 405.
Egami, F.: 1976, *Origins of Life* 7, 71.
Fripp, R. E. P., Donnelly, T. H. and Lambert, I. B.: 1978, *Spec. Publ. Geol. Soc. S. Afr.* 5, in press.
Garrels, R. M. and Mackenzie, F. T.: 1974, *Soc. Econ. Paleontol. Mineral. Spec. Publ.* 20, 193.
Goodwin, A., Monster, J. and Thode, H. G.: 1976, *Econ. Geol.* 71, 870.
Harrison, A. G. and Thode, H. G.: 1957, *Trans. Faraday Soc.* 53, 1648.
Hawkesworth, C. J.: Moorbath, S., O'Nions, R. K. and Wilson, J. F.: 1975, *Earth Planet. Sci. Lett.* 25, 251.
Heinrichs, T. K. and Reimer, T. O.: 1977, *Econ. Geol.* 72, 1426.
Holland, H. D.: 1973, *Geochim. Cosmochim. Acta* 37, 2605.
Holland, H. D.: 1974, *Soc. Econ. Paleontol. Mineral. Spec. Publ.* 20, 187.

Holser, W. T. and Kaplan, I. R.: 1966, *Chem. Geol.* 1, 93.
Hulston, J. R. and Thode, H. G.: 1965, *J. Geophys. Res.* 70, 3475.
Junge, C. E., Schidlowski, M., Eichmann, R. and Pietrek, H.: 1975, *Ibid.* 80, 4542.
Kaplan, I. R. and Rittenberg, S. C.: 1964, *J. Gen. Microbiol.* 34, 195.
Kemp, A. L. W. and Thode, H. G.: 1968, *Geochim. Cosmochim. Acta* 32, 71.
Knoll, A. and Barghoorn, E.: 1977, *Science* 198, 396.
Lambert, I. B.: 1978 *Publ. Geol. Dept. & Extension Service, Univ. West. Austr.* 2, 45.
Lambert, I. B., Donnelly, T. H., Dunlop, J. S. R. and Groves, D. I.: 1978, *Nature* 276, 808.
Lowe, D. R. and Knauth, L. P.: 1977, *J. Geol.* 85, 699.
Mäkelä, M.: 1974, *Geol. Surv. Finl. Bull.* 267, 45 p.
Monster, J., Appel, P. W. U., Thode, H. G., Schidlowski, M., Carmichael, C. M. and Bridgwater, D.: 1979, *Geochim. Cosmochim. Acta* 43, 405.
Moorbath, S., O'Nions, R. K. and Pankhurst, R. J.: 1973, *Nature* 245, 138.
Muir, M. and Grant, P. R.: 1976, In, *The Early History of the Earth* (B. F. Windley), Wiley, New York, pp. 595.
Peck, H. D.: 1974, In *Symp. Soc. Gen. Microbiol.* 24 (M. J. Carlile and J. J. Skehel, ed.), pp. 241.
Perry, E. C.: Monster, J. and Reimer, T.: 1971, *Science* 171, 1015.
Perry, E. C., Hickman, A. H. and Barnes, I. L.: 1975 *Geol. Soc. Amer. Abstr. Programs*, p. 1226.
Rye, R. O. and Ohmoto, H.: 1974, *Econ. Geol.* 69, 826.
Schidlowski, M.: 1973, *Geol. Rundsch.* 62, 840.
Schidlowski, M.: 1978, In *Origin of Life* (H. Noda, ed), Center Acad. Publ., Japan, p. 3.
Schidlowski, M., Appel, P. W. U., Eichmann, R. and Junge, C. E.: 1979, *Geochim. Cosmochim. Acta* 43, 189.
Schidlowski, M., Junge, C. E. and Pietrek, H.: 1977, *J. Geophys. Res.* 82, 2557.
Schneider, A.: 1970 *Contrib. Mineral. Petrol.* 25, 95.
Schopf, J. W.: 1974, In *Cosmochemical Evolution and the Origins of Life* (J. Oro, S. L. Miller, C. Ponnamperuma and R. S. Young, eds.), D. Reidel, Holland, p. 119.
Schwartz, R. M. and Dayhoff, M. O.: 1978, *Science* 199, 395.
Shima, M., Gross, W. H. and Thode, H. G.: 1963, *J. Geophys. Res.* 68, 2835.
Sidorenko, S. A. and Sidorenko, A. W.: 1975, *Geol. Inst. Acad. Nauk. SSSR Trudy* 277, 144 p.
Thode, H. G., Dunford, H. B. and Shima, M.: 1962, *Econ. Geol.* 57, 565.
Trudinger, P. A.: 1976, *Earth Sci. Rev.* 12, 259.
Vinogradov, V. I., Reimer, T. O., Leites, A. M. and Smelov, S. B.: 1976, *Litol. i Polezn. Iskop.* 11, 12.

STRUCTURAL FEATURES OF MANGANESE PRECIPITATING BACTERIA

KENNETH H. NEALSON and BRADLEY TEBO

Scripps Institution of Oceanography, A-002, La Jolla, Calif., 92093, U.S.A.

Abstract. Studies of biological communities of the past (and their associated activities) are usually dependent upon preservation of fossil material. With bacteria this rarely occurs because of the absence of sufficient fossilizable cellular material. However, some bacteria deposit metabolic products that can, conditions allowing, be preserved indefinitely. In particular, manganese and iron depositing bacteria have the capacity to form preservable microfossils. In order to better understand these microfossils of the past, we have examined present day morphologies of manganese oxidizing bacteria. These bacteria are highly pleomorphic, depending on the growth medium, the age of the culture, and the extent of manganese oxidation. Transmission electron microscopy indicates that manganese may be deposited either intra- or extra-cellularly. The prognosis of the use of morphological information for the interpretation of ancient and modern manganese deposits is discussed.

1. Introduction

Since the time of Beijerinck (1913) it has been known that bacteria can catalyze the oxidation of soluble divalent manganese (Mn^{2+}) to tetravalent manganese as the insoluble oxide (MnO_2), although exact mechanism remain unknown.

Pure cultures of a wide variety of bacterial taxa can be isolated from diverse environments: soils, lakes and oceans (Table I). The association of such manganese precipitating bacteria with actively accreting manganese precipitates has led many workers to hypothesize that bacterial activity is causally involved with the genesis and/or growth of the precipitates. Such hypotheses have been put forth for marine manganese nodules (Ehrlich, 1972, 1975), freshwater manganese deposits (Klaveness, 1977; see Kuznetzov, 1970 for review), water pipe deposits (Tyler and Marshall, 1967a, b, c) and soil precipitates (van Veen, 1973; Bromfield, 1956).

The role of microbiota in the formation of fossil manganese deposits is even more difficult to assess; any associated organisms are long since dead, making culture methods impossible and identification of microfossils very difficult. However, bacteria-like colonies have been identified in the iron facies of the remarkably well-preserved Gunflint chert (Barghoorn and Tyler, 1965). Correlation of laboratory and micropaleontological investigations are expected to eventually elucidate the origins of manganese precipitates. The product of the manganese precipitating bacteria is the insoluble oxide, MnO_2, a stable mineral under aerobic conditions. The interaction of these bacteria with sediments may leave a morphologically distinctive record in the form of a precipitate. If the precipitates

TABLE I
Bacterial genera that oxidize manganese[a]

Major group[b]	Class	Genus	Environment[c]	Reference
Pseudomonads		*Pseudomonas*	S	Zavarzin, (1962)
			M	Nealson (1978)
Facultative aerobic gram negative heterotrophs	Enterobacteria	*Achromobacter*	M	Ehrlich (1966)
		Aeromonas	M	Ehrlich (1966)
		Brevibacterium	M	Ehrlich (1966)
		Oceanospirillum	M	Ehrlich (1979)
		Vibrio	M	Ehrlich (1966)
		Arthrobacter (Siderocapsa)	M	Ehrlich (1966)
		Flavobacterium	M	Nealson (1978)
	Sphaerotilus group	*Leptothrix (Sphaerotilus)*	F	Mulder and van Veen (1963)
		Clonothrix	F	Wolfe (1960)
		Metallogenium	F	Kuznetzov (1970)
		Kuznetzovia	F	Kuznetzov (1970)
	Prosthecate bacteria	*Pedomicrobium*	S	Gebers and Hirsch (1978)
		Hyphomicrobium	F	Tyler and Marshall (1967a)
Gram positive aerobic endospore formers		*Micrococcus*	M	Ehrlich (1963)
		Bacillus	M	Ehrlich (1963)
Actinobacteria		*Nocardia*	S	Schweisfurth (1968)

[a] This table, not all inclusive, merely illustrates the diversity of manganese precipitating bacteria.
[b] Phyla and classes of Margulis (1974)..
[c] Isolated from marine (M), soil (S) or fresh water (F) environments.

are well-preserved, then morphological examination of fossil and live microbes and their precipitates may provide insight into processes of past and present metal precipitation. Clearly, however, a prerequisite is the identification of the organisms and their precipitates formed under controlled conditions. This chapter, which describes manganese precipitates formed in the laboratory as the result of bacterial activity, discusses the prognosis of the use of such morphological information to interpret ancient and modern MnO_2 deposits.

2. Materials and Methods

Manganese oxidizing bacteria were isolated from enrichment cultures as previously described (Nealson, 1978). They were grown and maintained on K medium: 2 g bactopeptone (Difco), 0.5 g yeast extract (Difco), 0.18 g $MnCl_2 \cdot 4H_2O$ and one liter of 75% seawater. The manganese chloride was made up as a 0.45% solution in distilled water, sterilized by filtration (0.2 μm Millipore membrane filter) and was added to the other components of the medium after they were autoclaved.

On solid medium (1.5% agar) and for qualitative manganese determinations, oxidation was detected by spot tests, using either Feigl's benzidinium reagent (1958), or leukoberbelin blue reagent (Krumbein and Altmann, 1973), both of which form characteristic blue-colored compounds upon reaction with manganese dioxide. The leukoberbelin blue (kindly supplied by Dr W. Ghiorse) offers the advantages that it is non-toxic, and reduces the oxidized manganese upon reaction, so that it can be used to gently remove MnO_2.

Samples were prepared for microscopy by fixation for up to three days with a 1% glutaraldehyde solution in sterile buffered seawater (75% seawater and 50 mM N-2-hydroxyethylpiperazine-N'-2-ethane sulfonic acid (HEPES) buffer). During fixation, samples were kept at 4° and then washed in sterile buffered medium at 4°.

For scanning electron microscopy, seawater was gradually substituted by distilled water (50, 75, and 100%). Samples were then dehydrated in ethanol (50, 60, 70, 80, 90, and 100%) for 10 min at each concentration. After dehydration, freon 113 was gradually substituted for ethanol. Samples were then dried by the critical point method and attached to aluminum stubs with either double-stick tape or carbon paint. Once mounted, the samples were sputter-coated with gold to a thickness of 100 Å with a Technics coater, studied and photographed with a Cambridge S-4 scanning electron microscope. Energy dispersive X-ray (EDS) analysis to determine elemental composition of the samples was performed at the same time, using an ORTEC computerized system.

Samples for transmission electron microscopy were fixed in 1% glutaraldehyde in artificial sea water (ASW: NaCl 200 mM; $MgSO_4 \cdot 7H_2O$, 50 mM; KCl, 10 mM; $CaCl_2 \cdot 2H_2O$, 10 mM) for up to three days, washed three times (15 min each) in ASW, and post-fixed in 1% osmium tetroxide (OsO_4) in ASW for 1–2 hr. Samples were then dehydrated in a graded series of ethanol solutions to 100%, and finally in 100% propylene oxide. They were transferred to Spurr low viscosity embedding media without accelerator and rotated on a turntable for up to 12 hr. Samples were then placed in BEEM capsules with fresh Spurr's medium and allowed to sit in a desiccator for 6–12 hr. Samples were then polymerized for 6 hr to 70°. Thin sections (60–80 Å) were cut on an LKB Ultramicrotome III and studied and photographed using a Zeiss Transmission electron microscope (Model 9AS) at an accelerating voltage of 75 keV.

3. Results and Discussion

The taxonomic, morphological, physiological and ecological diversity of manganese oxidizing bacteria is shown in Table I. Are the differences in manganese precipitates correlatable with differences in the associated bacteria? A goal is to distinguish deposits formed by different groups of bacteria for eventual identification of the organisms and processes that form naturally occurring precipitates.

Several manganese oxidizing bacteria grown in culture demonstrate features of bacterial manganese deposits in nature. Figure 1 shows a pure culture of a strain of bacillus, SG-1. Also shown is a spontaneous mutant (SG-1W) of the same strain incapable of manganese oxidation. A crusty brown precipitate develops on the surface of the colony of the man-

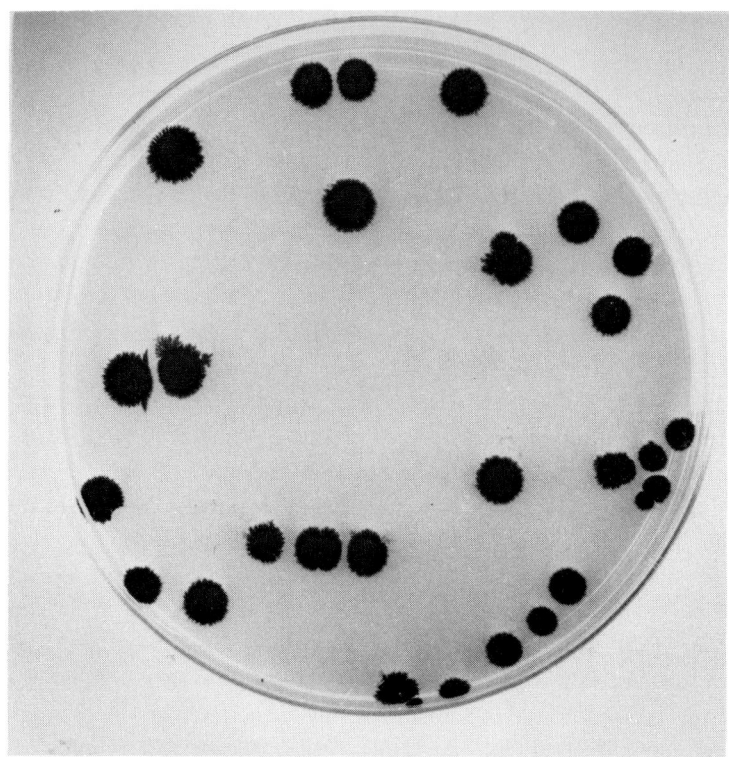
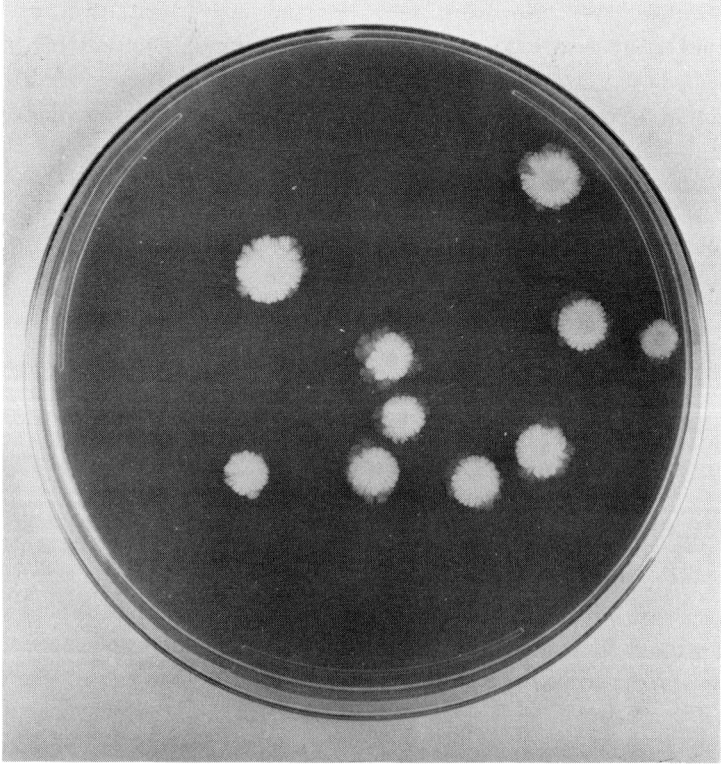

Fig. 1. A marine bacillus, strain SG-1 which rapidly oxidizes manganese (top) and its spontaneously derived mutant which is incapable of manganese oxidation (bottom). Two-week-old colonies; the black precipitate appears within a few days.

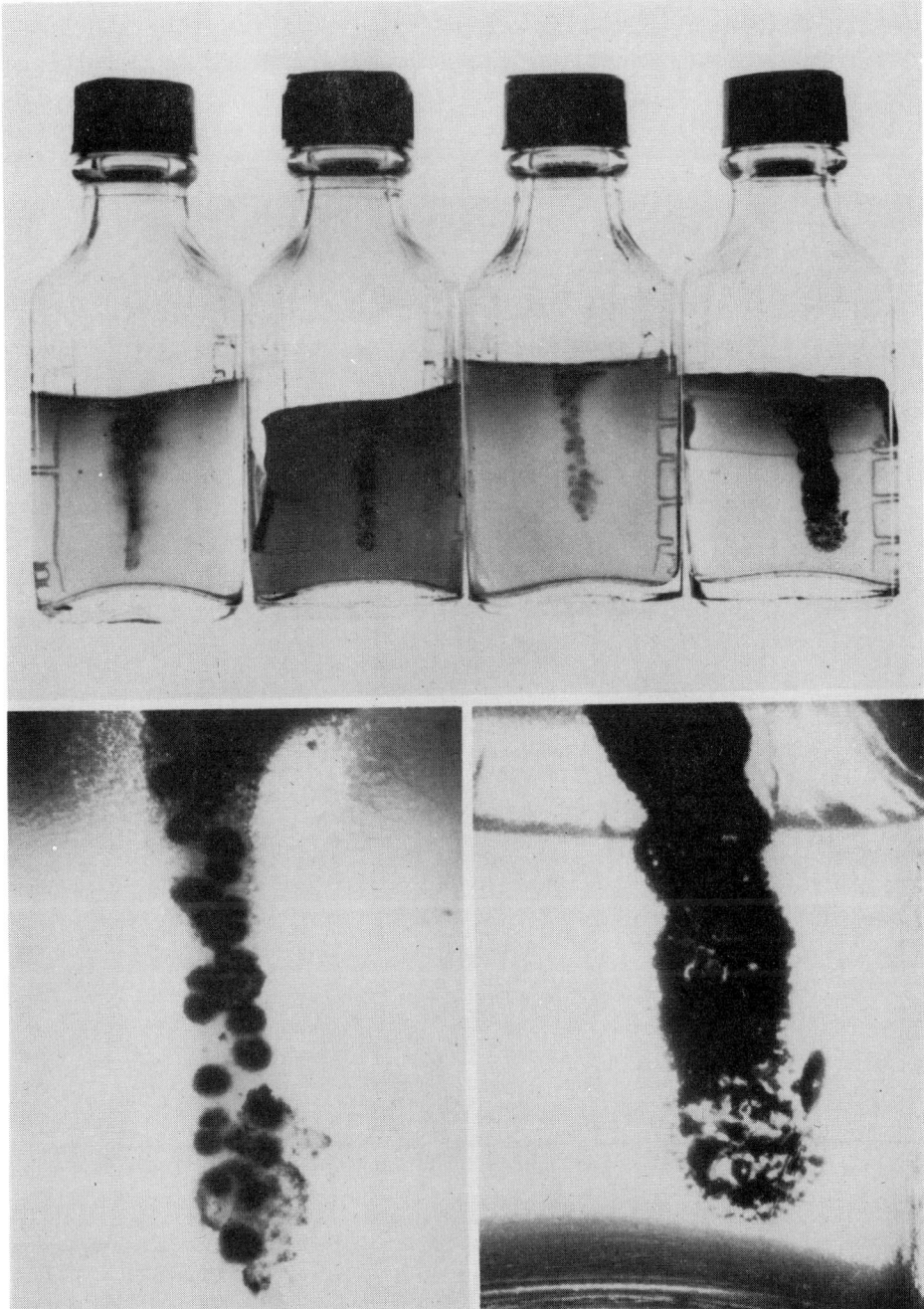

Fig. 2. Gram negative rod-shaped bacteria which oxidizes manganese slowly. These heavy precipitates took one year to develop in soft agar stab cultures. Manganese dioxide formation by these bacteria, which are probably microaerophilic, was originally detected on plates by benzidinium hydrochloride.

ganese oxidizing strain, while the mutant retains a characteristic white color, typical of many bacilli. However, for other bacteria, the low quantity and rate of manganese oxidation may preclude macroscopic precipitates; manganese oxidation is detected only when manganese test reagents such as benzidinium hydrochloride (Feigl, 1958) or leukoberbelin blue (Krumbein and Altmann, 1973) are used. After growth for at least one month,

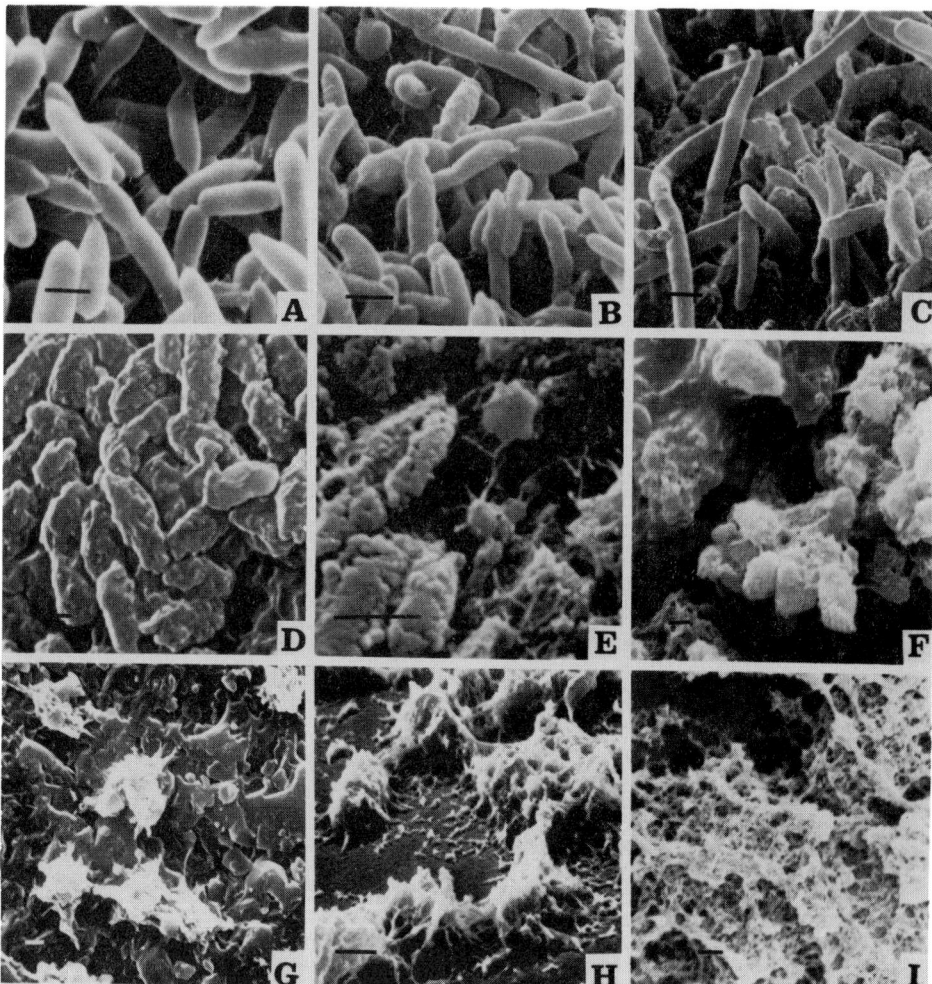

Fig. 3. Scanning electron micrographs of the same manganese oxidizing bacillus (strain SG-1) under different conditions. (Bars = 1.0 μm). (A) Agar medium (1.5%) lacking manganese, cells after one week of growth. (B) Cells after one day of growth in K medium (1.5% agar, 0.8 gm MnCl/liter). (C) Cells after one week of growth in K medium. (D) Cells taken from the top surface of the colony after one month of growth on K medium. (E) Cells taken from the bottom surface of the colony after one month of growth on K medium. (F) Clumps of manganese coated bacteria after several months of growth in manganese supplemented sea water. (G) Clumps of MnO_2 coated bacteria on sand grains in sea water after one week of growth. (H) Clumps of MnO_2 coated bacteria on glass slides after one week of growth. (I) Clumps of MnO_2 bacteria on glass slides after three months of growth.

however, nearly all manganese depositing bacteria become coated or associated with MnO_2. Figure 2 shows several cultures grown in stab cultures for one year. The subsurface precipitates are formed only when bacteria are present, and are associated with the bacterial cells.

The morphology of SG-1 cells depends on several factors: composition of the growth medium, the age of the culture, the extent of manganese precipitation and the amount of surface available. Some of the morphological variations are shown in Figure 3. On agar plates, colonies with no manganese in the growth medium (A) never form precipitates; on manganese-containing medium, young colonies (B) precipitate less manganese relative to old (C–E). There are also differences between the top surface precipitates (D) and the bottom surfaces (E) of the same colonies grown on agar plates. Cells attached to glass show variation with age (H, 1 week; I, 3 months); and cells clumped from liquid (F) or growing on a sand grain (G) are also distinctive. Strain SG-1 even includes many morphologically variant forms that are scarcely recognizable as bacteria (Figure 3E–I). The morphological variation of such bacterial precipitates must be understood in great detail if they are to be used to gain insight into past processes. All the different morphologies in Figure 3 are produced by only one species of bacteria. A similar range of variation in other manganese precipitating microbes is likely to be seen. In fact, a complex life cycle displaying many different morphologies is characteristic of the manganese oxidizing bacteria, *Metallogenium*. *Metallogenium* and its precipitates is at least as variable as SG-1, if not more so. This organism, which has been studied by several Russian scientists (for review, see Kuznetzov, 1970), appears to be the modern counterpart of the genus *Eoastrion* of Barghoorn and Tyler (Barghoorn, 1977). Variable morphology of the budding bacteria *Pedomicrobium* has also been reported by Ghiorse and Hirsch (1977) and Gebers and Hirsch (1977). Such heterogeneity impedes and in some cases may even preclude identification of manganese oxidizing organisms by examination limited to morphology of their precipitates. Furthermore, most MnO_2-associated microorganisms have not even been examined in this context.

Although there are difficulties as discussed, it is hoped that, after a catalog of structures is obtained, the identification of causative organisms will be possible. Furthermore, if identification is achieved, since the bacterial deposition of manganese varies as a function of growth conditions, it may eventually be possible to infer the environmental conditions existing at the time of metal deposition.

Thin sections of the subsurface deposits of Figure 2 examined by transmission electron microscopy show that the precipitates are obviously of bacterial origin (Figure 4). In one case, the manganese precipitation is entirely extracellular. Ghosts or cellular depressions similar to those observed might have been left in fossil material by such a process. In another example, the manganese is precipitated inside the cells rather than outside; fossils of these cells would certainly be quite different from those of the first type. The precipitates around spores of SG-1 are also extracellular, as shown in thin sections of SG-1 that were scraped from a glass surface (Figure 4E). Similar thin sections of strain SG-1, after the removal of precipitated manganese by treatment with leukoberbelin blue reagent, are shown in Figure 4F.

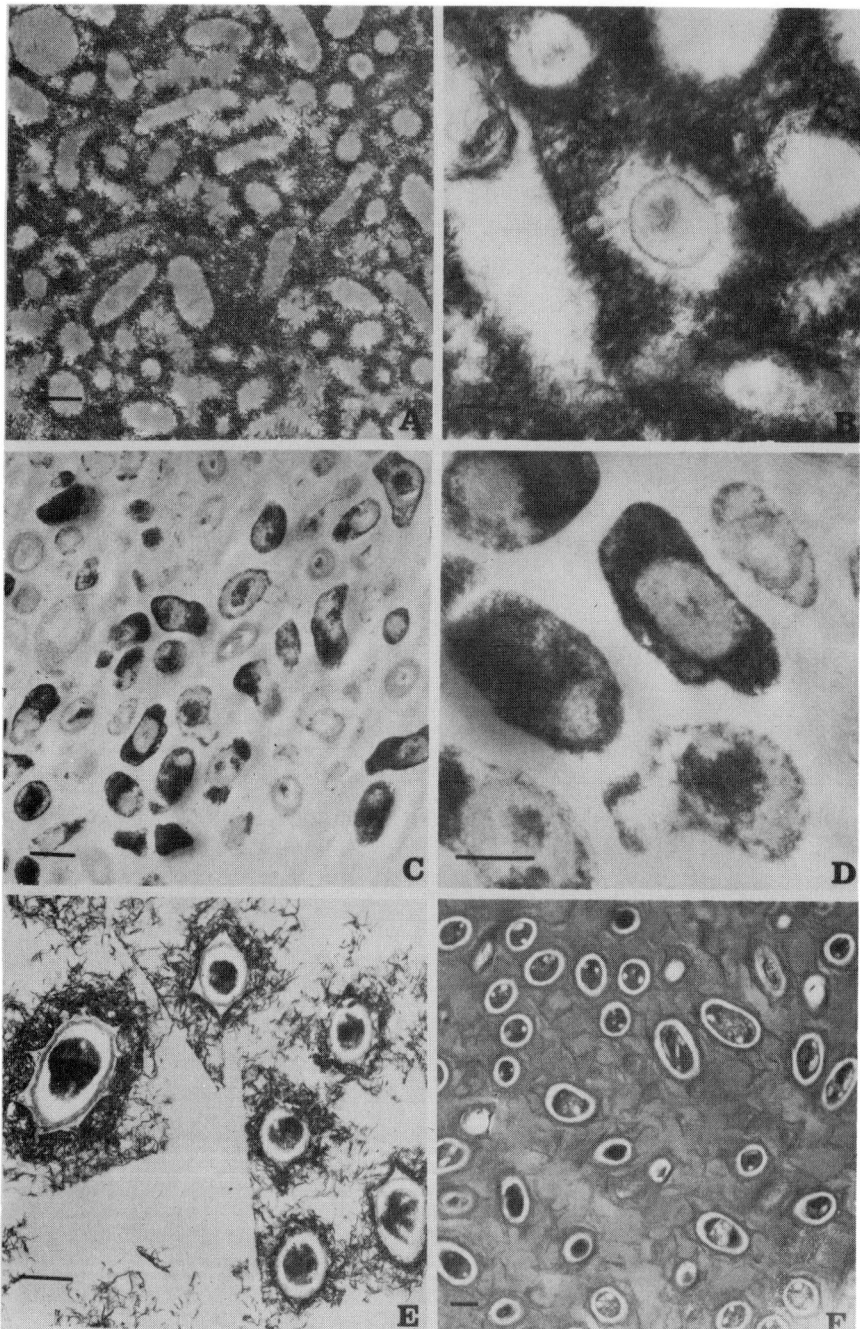

Fig. 4. Transmission electron micrographs of manganese precipitation by bacteria. (A) A gram negative rod (strain 56A) forms an extracellular precipitate of MnO_2. Bar = 0.5 μm. (B) Same as (A) but at higher magnification. Bar = 0.25 μm. (C) A second gram negative rod-shaped bacterium (strain 45B) that deposits MnO_2 inside the cells. Bar = 0.5 μm. (D) Same as (C) but higher magnification. Bar = 0.25 μm. (E) A third mode of MnO_2 deposition is seen in the bacillus (strain SG-1) in which the spores but not the cells become coated with MnO_2. Bar = 0.5 μm. (F) Spores of the bacillus like those in (E) which have been treated with leukoberbelin blue and the MnO_2 removed from the external surfaces of the spores Bar = 0.5 μm.

When precipitates are removed from the bacillus SG-1, characteristic bacterial structures remain (Nealson and Ford, 1980). Thus, with other manganese precipitates it may also be possible to identify organisms by removing the extracellular precipitates.

Further knowledge of the structure, metabolism, morphology and rates of production of precipitates of manganese oxidizing bacteria is prerequisite to understanding manganese accumulation in nature. Even after such information is available, determination of the origins of fossil and recent manganese precipitates may not be easy. Are bacterial precipitates really recognizable? How does the morphology of precipitates such as those seen in Figure 4 alter with continuing manganese deposition and diagenesis? Are unique or characteristic oxides or mineral phases produced by bacteria? What is their preservation potential? Are some precipitates only biogenic? Do certain oxides or mineral phases require conditions of temperature and pressure that preclude their biogenicity? Answers to these questions are intrinsic to an understanding of the biology and geology of the Precambrian environment in general, as well as to that of recent manganese concretions.

Acknowledgements

This contribution is a part of the MANOP program of the National Science Foundation, and the research was supported by NSF-OCE-7827383 to K. H. Nealson. We are grateful to W. S. Moore and Jeanne Ford for aid with both the work and the manuscript.

References

Barghoorn, E. S.: 1977, in *Chemical Evolution of the Early PreCambrian*, Academic Press, New York, pp. 185–186.
Barghoorn, E. S. and Tyler, S. A.: 1965, *Science* 147, 563.
Beijerinck, M. W.: 1913, *Folia Microbiol. (Delft)* 2, 123.
Bromfield, S. M.: 1956, *Australian J. Biol. Sci.* 9, 238.
Dubinina, G. A., Gorlenko, V. M., and Suliemanov, Ya. I.: 1974, *Microbiol.* 42, 817.
Ehrlich, H. L.: 1963, *Appl. Microbiol.* 11, 15.
Ehrlich, H. L.: 1966, *Dev. Ind. Microbiol.* 7, 279.
Ehrlich, H. L.: 1972, in D. R. Horn (ed.), *Ferromanganese Deposits on the Ocean Floor*, National Science Foundation, Washington, D.C., pp. 63–70.
Ehrlich, H. L.: 1975, *Soil Sci.* 119, 36.
Ehrlich, H. L. and Arcuri, E. J.: 1979, *Proceedings American Society of Microbiology* 66, 190.
Feigl, F.: 1958, *Spot Tests in Inorganic Analysis*, Elsevier, New York.
Gebers, R. and Hirsch, P.: 1978, in W. E. Krumbein (ed.), *Environmental Biogeochemistry and Geomicrobiology*, Vol. 3, Ann Arbor Science, Ann Arbor, Michigan, pp. 911–922.
Ghiorse, W. C. and Hirsch, P.: 1978, in W. E. Krumbein (ed.), *Environmental Biogeochemistry and Geomicrobiology*, Vol. 3, Ann Arbor Science, Ann Arbor, Michigan, pp. 897–909.
Klaveness, D.: 1977, *Hydrobiol.* 56, 25.
Krumbein, W. E. and Altmann, H. J.: 1973, *Helgol. Wiss. Meeresuntees* 25, 347.
Kuznetzov, S. I.: 1970, in C. H. Oppenheimer (ed.), *The Microflora of Lakes and its Geochemical Activity*. University of Texas Press, Austin Texas.
Margulis, L.: 1974, *Handbook of Genetics* 1, 1.
Marshall, K. C.: 1979, in P. A. Trudinger and D. J. Swaine (eds.), *Biogeochemical Cycling of Mineral-Forming Elements*, Elsevier, Amsterdam, pp. 253–292.

Mulder, E. G. and van Veen, W. L.: 1963, *J. Microbiol. Serol.* **29**, 121.
Nealson, K. H.: 1978, in W. E. Krumbein (ed.), *Environmental Biogeochemistry and Geomicrobiology*, Vol. 3, Ann Arbor Science, Ann Arbor, Michigan, pp. 847–858.
Nealson, K. H. and Ford, J.: 1980, *Geomicrobiol. J.* **2**, 21.
Schweisfurth, R.: 1968, *Mitt. Int. Ver. Theor. Angew. Limnol.* **14**, 179.
Tyler, P. A. and Marshall, K. C.: 1967a, *Arch. Mikrobiol.* **56**, 344.
Tyler, P. A. and Marshall, K. C.: 1967b, *J. Am. Water Works Assoc.* **59**, 1043.
Tyler, P. A. and Marshall, K. C.: 1967c, *Antonie van Leeuwenhoek* **33**, 171.
van Veen, W. L.: 1973, *Antonie van Leeuwenhoek* **39**, 657.
Wolfe, R. S.: 1960, *J. Am. Water Works Assoc.* **52**, 1335.
Zavarzin, G. A.: 1962, *Microbiol.* **31**, 481.

THE RADIORACEMIZATION OF AMINO ACIDS BY IONIZING RADIATION: GEOCHEMICAL AND COSMOCHEMICAL IMPLICATIONS*

WILLIAM A. BONNER and NEAL E. BLAIR

Department of Chemistry, Stanford University, Stanford, Calif. 94305, U.S.A.

and

RICHARD M. LEMMON

Lawrence Berkeley Laboratory, University of California, Berkeley, Calif. 94720, U.S.A.

Abstract. A number of optically active amino acids, both in the solid state and as sodium or hydrochloride salts in aqueous solution, have been exposed to ionizing radiation from a 3000 Ci ^{60}Co γ-ray source to see if radioracemization might accompany their well-known radiolysis. γ-Ray doses causing 55–68% radiolysis of solid amino acids typically engendered 2–5% racemization, while aqueous solutions of the sodium salts of amino acids which underwent 53–66% radiolysis showed 5–11% racemization. Amino acid hydrochloride salts in aqueous solution, on the other hand, showed little or no radioracemization accompanying their radiolysis. Both radiolysis and radioracemization were roughly proportional to γ-ray dose in the range studied ($1-36 \times 10^6$ rads). Mechanisms for the radioracemization of amino acids in the solid state and as aqueous sodium salts are discussed, and the absence of radioracemization for aqueous hydrochloride salts is rationalized. Isovaline, a non-protein amino acid which has been isolated from the Murchison meteorite, contains no α-hydrogen atom and is therefore incapable of racemization *via* the chemical mechanisms by which ordinary amino acids racemize. Nevertheless, isovaline suffers radioracemization in the solid state to an extent comparable to that shown by ordinary amino acids, as do its sodium and hydrochloride salts in the solid state. The sodium salt of isovaline in aqueous solution, however, fails to racemize during its radiolysis. Several implications of the newly described phenomenon of radiomization are pointed out for the fields of geochemistry and cosmochemistry.

1. Introduction

In recent years we have been involved in a variety of experimental investigations (Bonner, 1974; Bonner and Flores, 1975, Bonner *et al.*, 1975, 1976/77, 1978; Noyes *et al.*, 1977) probing the possible validity of the Vester-Ulbricht mechanism (Vester *et al.*, 1959; Ulbricht, 1959; Ulbricht and Vester, 1962) for the origin of optical activity in nature *via* parity violation during the β-decay or radionuclides. In one series of such experiments (Bonner *et al.*, 1978) a number of 17–25 year-old samples of crystalline ^{14}C-labelled amino acids, both racemic and optically active, of high specific radioactivity were examined by gas chromatography (G.C.) for both gross radiolysis and possible stereoselective (asymmetric) degradation. These samples, which had undergone self β-radio-

---

*A portion of this research has been described previously at the 144th National Meeting of the American Association for the Advancement of Science, Washington D.C., Feb. 12–17, 1978 and at the Carnegie Institution of Washington Conference: Advances in the Biogeochemistry of Amino Acids, Airlie House, Warrenton, Virginia, Oct. 29 – Nov. 1, 1978.

lysis during the several decades since their original preparation and purification (Bernstein et al., 1972), showed as much as 67% gross degradation, but no asymmetric radiolysis. Several of the optically active ^{14}C-labelled amino acids, however, gave G.C. analyses which suggested that radiation induced racemization might also have accompanied their gross radiolysis. Since the possibility of such *radioracemization* had not been considered in any of the above (or other) experiments involving attempted stereoselective β-radiolysis, and since such a possibility had been only scarcely suggested in earlier literature (Feng and Tobey, 1959; Evans, 1966; Evans et al., 1968), we decided to undertake a systematic preliminary investigation of the phenomenon. Our general plan was to subject a number of crystalline or dissolved optically pure D- and L- amino acids to heavy doses of ionizing radiation, then to examine the partially destroyed residue by G.C. for both gross radiolysis and possible radioracemization.

2. Experimental

The samples irradiated consisted of 10–17 mg portions, either solid or dissolved, of the optically pure (as determined by G.C.) amino acids listed in the Tables below. The samples were of the purest quality available from Aldrich Chemical Co., Calbiochem, Mann Research Laboratory, Nutritional Biochemicals, or Sigma Chemical Co. For hydrochloride salt irradiations the weighed amino acid was first dissolved in an equivalent amount of 0.1 N aqueous HCl, while for sodium salt irradiations each sample was dissolved in one equivalent volume of 0.1 N NaOH. The solid or dissolved samples were placed in 1 x 4 cm glass vials stoppered with Bakelite caps fitted with Teflon gaskets. The vials were placed in a 3000-Ci ^{60}Co γ-ray source at the Lawrence Berkeley Laboratory, a source designed to deliver high dose rates (4–10 x 10^6 rads hr^{-1}) to small samples. Irradiations proceeded for time periods (5–90 hrs.) sufficient to afford the total radiation doses shown in the following Tables and to provide *ca*. 50–70% gross radiolysis.

After irradiation, each solid sample was dissolved and each liquid sample diluted to a volume of 5.0 ml with water or dilute HCl, then was quantitatively divided in half volumetrically, and all solutions were finally rotary-evaporated to dryness under vacuum. One of the 50% aliquots of each sample was then treated with a weighed quantity of the corresponding enantiomeric (or racemic) amino acid, so as to permit determination of the percent degradation of the sample by the 'enantiomeric marker' technique (Bonner, 1973). The amino acid samples were then converted into their *N*-trifluoroacetyl (*N*-TFA) isopropyl ester derivatives (I) for G.C. analysis (Bonner et al., 1974), first by esterifying with 2-PrOH/HCl, the acylating with trifluoroacetic anhydride. In the case of the G.C. analyses of irradiated isovaline samples, however, enantiomeric derivatives of the type I

$$\begin{array}{cc} \text{R} \\ | \\ \text{F}_3\text{CCONHCHCOOCH(CH}_3\text{)}_2 \end{array} \qquad \begin{array}{c} \text{CH}_3 \quad \text{CH}_2\text{CH(CH}_3\text{)}_2 \\ | \qquad \qquad | \\ \text{F}_3\text{CCONH—C—CONHCHCOOCH(CH}_3\text{)}_2 \\ | \\ \text{CH}_2\text{CH}_3 \end{array}$$

I II

proved unsatisfactory in their resolvability characteristics. Suitable quantitative analyses (i.e. baseline resolution) could be achieved, however, by converting the isovaline first into its diastereomeric N-TFA-isovalyl-D(or -L)-leucine isopropyl ester derivatives (II) (Flores et al., 1977; Bonner et al., 1979b).

Each derivatized sample was then quantitatively analyzed by G.C. for its enantiomeric composition, using 46 m x 0.5 mm stainless steel capillary columns coated with one or the other of the two optically active enantiomeric G.C. phases, N-docosanoyl-D(or -L)-valine *tert*-butylamide (Bonner and Blair, 1979). The columns were installed in a Hewlett-Packard 5700A gas chromatograph and peak area integration was accomplished with the aid of a Hewlett-Packard 3380A digital electronic integrator-recorder, affording the analytical reproducibility and precision previously described (Bonner et al., 1974).

In order to assess the effect of increasing radiation dosage on the extents of both gross radiolysis and racemization, a number of D- or L- leucine samples (15–17 mg) were dissolved in the equivalent volumes of 0.1 N HCl or NaOH, and the solutions were irradiated in the above γ-ray source for increasing time periods, so as to achieve the increasing dosages shown in Table IV. The data in the Tables below have been previously presented separately in several publications (Bonner and Lemmon, 1978a, 1978b; Bonner et al., 1979a, 1979b).

3. Results

Tables I and II indicate that γ-radiation causes not only the previously documented radiolysis of amino acids (Garrison, 1968, 1972), but also engenders significant radioracemization as well – both in the solid state and as sodium salts in aqueous solution. Extensive radiolysis also occurs with the dissolved hydrochloride salts of amino acids (Table III) but, in contrast to the sodium salts (Table II), little or no concomittant

TABLE I
γ-Radiolysis and radioracemization of solid amino acids

No. Amino acid	Radiation dose (Rads x 10^{-8})	Decomposition, %	Residual amino acid		
			%D[a]	%L[a]	Racemization, %[b]
1. L-Alanine	8.1	38.6	1.9	98.1	3.8
2. D-2-Aminobutyric acid	8.1	55.8	99.2	0.8	1.6
3. L-Norvaline	8.1	66.1	1.6	98.4	3.2
4. L-Norleucine	8.1	63.1	1.3	98.7	2.6
5. D-Leucine	8.1	67.9	97.2	2.8	5.6
6. L-Leucine	8.1	68.0	2.5	97.5	5.0
7. D-Leucine	10.2	96.1	96.2	3.8	7.6
8. L-Leucine	10.2	93.2	6.8	93.2	13.6

[a]Standard deviations for averages of 2–4 G.C. analyses were generally ± 0.1–0.3%. Same in other Tables.
[b]2-Times the percent of inverted enantiomer.

TABLE II
γ-Radiolysis and radioracemization of amino acid sodium salts (0.10 M aqueous solution)

No.	Amino acid	Radiation dose (Rads x 10^{-7})	Decomposition, %	Residual amino acid		
				%D	%L	Racemization, %
1.	L-Alanine	1.7	65.8	5.8	94.2	11.6
2.	D-2-Aminobutyric acid	1.8	58.0	96.0	4.0	8.0
3.	D-2-Aminobutyric acid	1.8	59.4	95.8	4.2	8.4
4.	L-Norvaline	1.7	59.6	4.2	95.8	8.4
5.	L-Norleucine	1.7	52.8	5.3	94.7	10.6
6.	L-Valine	1.7	55.3	5.5	94.5	11.0
7.	L-Leucine	1.7	54.4	2.6	97.4	5.2

racemization takes place. Table IV indicates that increasing radiation dosage causes increasing radiolysis of leucine salts in aqueous solution as well as increasing radioracemization of the sodium salt. Again the hydrochloride salt samples underwent little or no racemization. A plot (see Figure 1) of the data in Table IV shows that radiolysis of the sodium salt of leucine occurs slightly faster than radiolysis of its hydrochloride salt, and that for both salts the extent of radiolysis is roughly proportional to total dosage up to *ca.* 60% degradation. Tolbert et al. (1962) have found a similar initial linear relation between radiolysis and dose for crystalline amino acids, and we have previously noted a linear decomposition-dose relationship during the first 80% of the γ-radiolysis of solid leucine (Bonner, 1973). Figure 1 also indicates an approximately linear relationship between extent of radioracemization and dosage for the sodium salt of leucine within the dose range studied. A rough indication of the overall reproducibility of the radiolysis and racemization data in the above Tables is seen in Nos. 2 and 3 of Table II, duplicate experiments in which identically sized samples of the dissolved sodium salt of 2-aminobutyric acid were irradiated simultaneously.

For reasons discussed below, it became desirable to investigate the γ-radiolysis and possible radioracemization of isovaline (α-amino-α-methyl-butyric acid, III), and to com-

TABLE III
γ-Radiolysis and radioracemization of amino acid hydrochloride salts (0.10 M aqueous solution)

No.	Amino acid	Radiation dose (Rads x 10^{-7})	Decomposition, %	Residual amino acid		
				%D	%L	Racemization, %
1.	L-Alanine	2.2	53.4	0.2	99.8	0.4
2.	D-2-Aminobutyric acid	2.2	52.1	100.0	0.0	0.0
3.	L-Norvaline	2.2	57.9	0.0	100.0	0.0
4.	L-Norleucine	2.2	63.5	0.0	100.0	0.0
5.	L-Valine	2.2	54.8	0.0	100.0	0.0
6.	L-Leucine	2.2	55.3	0.0	100.0	0.0

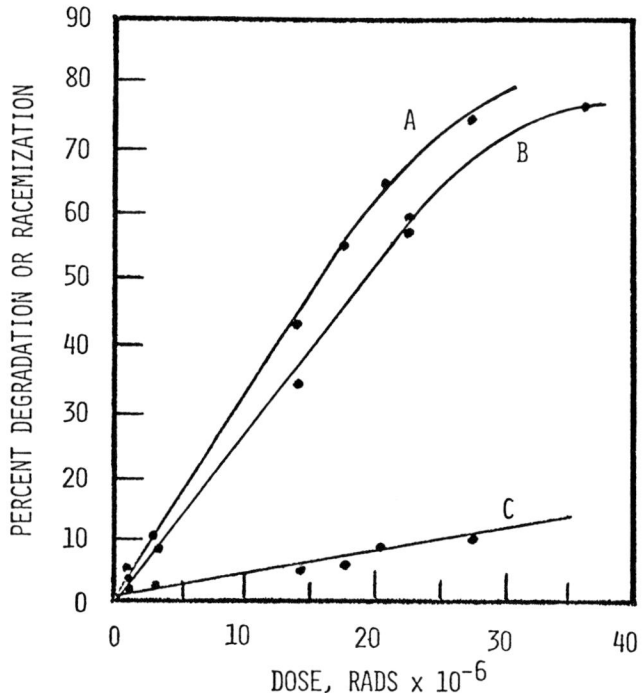

Fig. 1. Radiolysis and radioracemization of leucine salts (0.1M aqueous solution) by γ-radiation. (A) Degradation, Na Salt. (B) Degradation, HCl Salt. (C) Racemization, Na Salt.

TABLE IV
γ-Radiolysis and radioracemization of leucine salts (0.10 M aqueous solution) as a function of γ-ray dosage

Enantiomer	Salt	Radiation dose (Rads x 10^{-6})	Decomposition, %	Residual leucine		Racemization, %
				%D	%L	
D	Na	1.0	1.8	99.2	0.8	1.6
D	Na	3.0	9.4	99.2	0.8	1.6
D	Na	14.0	41.7	97.9	2.1	4.2
L	Na	17.3	54.4	2.6	97.4	5.2
D	Na	20.5	63.1	96.2	3.8	7.6
D	Na	27.0	73.6	95.4	4.6	9.2
D	HCl	1.0	3.3	99.8	0.2	0.4
D	HCl	3.0	7.6	99.7	0.3	0.6
D	HCl	14.0	32.5	99.2	0.8	1.6
L	HCl	22.0	55.3	0.0	100.0	0.0
D	HCl	22.1	57.6	—[a]	—	—
D	HCl	36.1	75.7	—[a]	—	—

[a] G.C. traces obscured by peaks for degradation products.

TABLE V
γ-Radiolysis and radioracemization of isovaline

No. Substrate	Radiation dose (Rads x 10^{-8})	Degradation, %	Racemization, %
1. D-Ival[a]	9.0	79.6	4.9
2. L-Ival[a]	9.0	78.0	4.7
3. D-Ival[a]	12.0	86.5	6.1
4. L-Ival[a]	12.0	86.8	6.4
5. D-Ival · H_2O[b]	9.0	80.0	2.5
6. L-Ival · H_2O[b]	9.0	79.0	3.2
7. D-Ival Na salt[a]	9.0	87.7	3.8
8. L-Ival HCl salt[a]	9.0	91.8	3.3
9. D-Ival Na salt (aq)[c]	0.25	68.1	0.0
10. L-Ival Na salt (aq)[c]	0.25	68.8	0.0

[a] Solid, anhydrous
[b] Monohydrate crystals
[c] 0.1 M aqueous solution

pare these with the results for the ordinary amino acids in Tables I-III. The principal

$$CH_3CH_2-\underset{\underset{NH_2}{|}}{\overset{\overset{CH_3}{|}}{C}}-COOH$$

III

results of these experiments are summarized in Table V, where we observe that isovaline samples in the solid state, either hydrated or anhydrous, or as the solid sodium (No. 7) or hydrochloride salt (No. 8), undergo significant radioracemization accompanying their γ-radiolysis. Again (Nos. 3,4 versus 1,2) as with solid (Table I) or dissolved (Table IV) leucine, increasing radiation doses cause an increase in both radiolysis and racemization. Isovaline in the hydrated crystal form (Nos. 5,6) appears slightly less susceptible to radioracemization than does the anhydrous form (Nos. 1,2) for the same radiation dose, though the two forms are approximately equally prone to radiolysis. While the radiolyses and radioracemizations of isovaline in the solid state are roughly comparable to those of the other amino acids in Table I, the complete absence of radioracemization for the aqueous sodium salt of isovaline (Nos. 9,10) is noteworthy in its contrast to the facile racemizations summarized in Table II. This difference is discussed below. Again, the approximate reproducibility of our radiolysis and radioacemization data are indicated by the effects noted at equivalent dosages for the various enantiomeric pairs in Table V (Nos. 1,2; 3,4; 5,6; 9,10).

4. Mechanistic Speculations

At the present time we have not experimentally established any mechanisms whereby the above radioracemization of solid amino acids or their aqueous sodium salts might occur, nor have we characterized the radiolysis products responsible for the several extraneous peaks in the various G.C. traces of our enantiomeric composition analyses. One can presently only speculate on mechanisms for radioracemization as they might be superimposed upon currently accepted mechanisms for the radiolysis of solid amino acids or their aqueous solutions. Garrison (1968, 1972) has recently discussed such mechanisms in detail.

The initial step in the γ-radiolysis of a solid amino acid zwitterion (IV) is thought to be carbon-hydrogen scission at the α-carbon atom producing an α-radical (V), a proton, and a secondary electron, e^-_s (Eq. (1)). The secondary electron as well as the α-radical V then attack both IV and other intermediate species trapped in the crystal lattice,

$$\underset{IV}{\underset{R-CH-COO^-}{\overset{+NH_3}{|}}} \xrightarrow{\gamma\text{-photon}} \underset{V}{\underset{R-\underset{\cdot}{C}-COO^-}{\overset{+NH_3}{|}}} + H^+ + e^-_s \qquad (1)$$

ultimately affording the observed radiolysis products. In such a reaction sequence, to the extent that the α-radical V abstracts hydrogen atoms from other constituents of the crystal lattice to re-form IV, the regenerated IV should be totally or partially racemic, since V is presumably unable fully to maintain its original stereochemical configuration. Clearly, some similar sequence of degradative reactions, one or more of which regenerates the original but racemized amino acid, must apply also to the radioracemization of solid isovaline. Since isovaline has no α-hydrogen, however, the initial radiation-induced α-carbon scission must involve a C-C rather than a C-H band.

The radiolysis of amino acids in aqueous solution is believed (Garrison, 1968, 1972) to be initiated by hydrated electrons, e^-_{aq}, or hydroxyl radicals, HO·, produced by prior radiolysis of the water solvent (Eq. (2)). These species in turn attack the amino acid and

$$H_2O \xrightarrow{\gamma\text{-photon}} e^-_{aq} + HO\cdot + H^+ + H_2 + H_2O_2 \qquad (2)$$

initiate a series of degradation reactions similar to those postulated for solid state radiolysis, ultimately affording the observed fatty acid and α-keto acid degradation products. In particular, α-hydrogen abstraction by HO· can produce a key radiolysis intermediate, α-radical V (Eq. (3)). Reversal of reaction (3) would clearly provide a facile mechanism for the radioracemization of zwitterion IV in aqueous solution.

$$HO\cdot + \underset{IV}{\underset{R-CH-COO^-}{\overset{+NH_3}{|}}} \longrightarrow H_2O + \underset{V}{\underset{R-\underset{\cdot}{C}-COO^-}{\overset{+NH_3}{|}}} \qquad (3)$$

Our solution radiolysis, however, were conducted using sodium (VI) or hydrochloride salts (VII) of the amino acids in question. Here the mechanism for radioracemization

$$\underset{\text{VI}}{\text{R}-\overset{\overset{\text{NH}_2}{|}}{\text{CH}}-\text{COO}^-, \text{Na}^+} \qquad \underset{\text{VII}}{\text{R}-\overset{\overset{^+\text{NH}_3, \text{Cl}^-}{|}}{\text{CH}}-\text{COOH}}$$

would again presumably be HO· attack to form a stereochemically labile α-radical analogous to V, followed by solvent attack to regenerate racemized IV or VII, akin to the reversal of Eq. (3). However, our findings were that the sodium salts VI were particularly susceptible to radioracemization, while the hydrochloride salts VII under similar conditions were essentially unracemized. The facile radioracemization of the aqueous sodium salts of amino acids has been rationalized (Bonner and Lemmon, 1978b) as resulting from the ready formation of the α-radical from the anion of salt VI, due to its stabilization as the highly symmetrical resonance hybrid VIII. Such a resonance-stabilized intermediate cannot arise from the corresponding α-radical (IX) from the cation of VII, thus

$$\left[\text{R}-\overset{\overset{\text{NH}_2}{|}}{\underset{\cdot}{\text{C}}}-\text{C}\overset{\cdot\cdot\overset{\cdot\cdot}{\text{O}}\cdot}{\underset{\cdot\overset{\cdot\cdot}{\text{O}}\cdot}{\diagup}}\right]^- \longleftrightarrow \left[\text{R}-\overset{\overset{\text{NH}_2}{|}}{\text{C}}=\text{C}\overset{\cdot\cdot\overset{\cdot\cdot}{\text{O}}\cdot}{\underset{\cdot\overset{\cdot\cdot}{\text{O}}\cdot}{\diagup}}\right]^{--} \qquad \text{R}-\overset{\overset{^+\text{NH}_3}{|}}{\underset{\cdot}{\text{C}}}-\text{C}\overset{\text{O}}{\underset{\text{OH}}{\diagup}}$$

VIII   IX

explaining its apparent slow formation and the lack of accompanying radioracemization. Other speculations regarding the radioracemization of VI and the optical stability of VII have been advanced by Bonner and Lemmon (1978b).

The complete absence of radioracemization for the sodium salt of aqueous isovaline (Table V) is noteworthy in its contrast to the extensive racemization of the sodium salts of common amino acids (Table II). This observation is understandable in terms of Eq. (3). If the radioracemization of the sodium salts of amino acids is intitiated by analogous α-hydrogen abstraction by HO· to form the resonance stabilized α-radical VIII, such a mechanism is not available to isovaline as it lacks the requisite α-hydrogen atom. The sodium salt of isovaline is thus immune to radioracemization in aqueous solution (but not in the solid state), although gross radiolysis (by other reaction paths) is extensive (Table V).

5. Geochemical and Cosmochemical Implications

The racemization of natural amino acids under diagenetic environmental conditions is a phenomenon of considerable current interest to geochemists, paleontologists and archeologists, since it has been widely assumed that the D/L ratios for residual amino acids isolated from ancient specimens may provide a measure of the age of the specimen. Assuming that the amino acids isolated from a prehistoric sample are of protein origin and were therefore originally of the L-configuration and that the simple first order kinetics applicable to the racemization of amino acids in solution are similarly valid for the natural diagenetic racemization of these amino acids, it is possible — after establishing the D/L ratios of several amino acids in a sample of known age — to calculate the specific rate constants for the diagenetic racemization of those amino acids. Assuming also a uniform external environment at the given sample site within the geological epoch in question, one can then use the calculated rate constants to estimate the ages of other prehistoric samples from the same general area, again by determining the D/L ratios of the same amino acids isolated from the latter samples. In this general way D/L enantiomer ratios have been utilized during the past decade or so for the age dating of ancient specimens of geological sediments (Bada et al., 1970; Wehmiller and Hare, 1971), shells (Hare and Mitterer, 1968), bones (Bada, 1972; Bada and Protsch, 1973; Dungworth et al., 1974), teeth (Helfman and Bada, 1975, 1976), and corals (Wehmiller et al., 1976). In addition, having once independently established the age of a particular prehistoric bone sample (e.g., by ^{14}C dating), and having then also determined in the laboratory the temperature effects on the rates at which certain amino acids in modern bones racemize or epimerize, several investigators have thereupon used subsequently determined enantiomer or epimer ratios of these amino acids in the sample to estimate the temperatures prevailing over the past geological lifetime of the sample i.e., as paleotemperature indicators (Bada et al., 1973; Schroeder and Bada, 1973; Mitterer, 1975).

The validity of applying amino acid racemization criteria to geochronology and geothermometry has recently been critically questioned (Dungworth, 1976; Williams and Smith, 1977), and a host of previously ignored environmental and other factors have been enumerated which could cause serious errors in estimates of the rate constants for the diagenetic racemization of amino acids. Such errors in turn could seriously invalidate geochronological or geothermometric conclusions based on the simple measurement of D/L enantiomer ratios for the amino acids in question. To these formidable pitfalls challenging the validity of D/L ratios alone as criteria for estimating geological ages or temperatures must now be added the possibility of radioracemization during the epoch in question. While uniform radioactivity in the earth's crust might conceivably contribute relatively uniformly to the diagenetic racemization of amino acids in ancient samples, the proximity of additional radioactive material — either in an indigenous mineral matrix or dissolved in ground waters — would clearly induce additional indeterminate amounts of radioracemization in the samples in question, and thus suggest their spurious antiquity. Furthermore — a question which we are now investigating experimentally — it seems possible that clay minerals (in conjunction with radioactive sources) might enhance the effectiveness of radioracemization, as they apparently do for thermal racemization (Kroepelin, 1968; Flores and Bonner, 1974), thus shortening further the time required for the diagenetic racemization of ancient amino acid samples. In summary we would argue that unless specific knowledge of the radiation exposure history of a given ancient amino acid sample is available, the phenomenon of radioracemization might unwittingly cause considerable error in geochronological or geothermometric conclusions based solely on the measurement of the simple D/L enantiomer (or epimer) ratio of the amino acid in question.

Our subsequent interest (Bonner *et al.*, 1979a, 1979b) in the possible radio-racemization of isovaline (III) was occasioned by the cosmological importance of this non-protein amino acid. In September 1969, a type II carbonaceous chondrite fell to earth near Murchison, Victoria, Australia, and shortly thereafter extensive investigations were undertaken into the organic constituents of this fragment, using G.C. and other analytical techniques (Kvenvolden *et al.*, 1970, 1971; Oró *et al.*, 1971a, 1971b; Cronin and Moore, 1971; Lawless, 1973; Lawless and Peterson, 1975). As early as 1971 Kvenvolden *et al.* reported 18 amino acids to be present in the Murchison chondrite, 12 of which including isovaline, were non-protein and 6 of which were common to terrestrial proteins, and by 1973 Lawless had extended the number of amino acids present to 45. During their G.C. analyses, the optically active amino acids from the Murchison meteorite were found to consist of approximately equal amounts of D- and L-enantiomers (Kvenvolden *et al.*, 1971, 1972; Oró *et al.*, 1971b), a fact interpreted as indicating their probable abiotic origin. The presence of non protein amino acids in the Murchison and Murray meteorites has led to the same interpretation (Kvenvolden *et al.*, 1971; Lawless *et al.*, 1971).

Of the non-protein amino acids from the Murchison chondrite isovaline has been of particular interest cosmologically since, in contrast to other amino acids thus far isolated, it lacks a hydrogen atom on its α-carbon atom. Accordingly, isovaline cannot undergo racemization by the known mechanisms responsible for the racemization of common amino acids, namely, reactions involving reversible scission of the α-C-H bond (Pollock *et al.*, 1975). For this reason, it has been argued (Lawless, 1973) that the enantiomeric composition of the Murchison isovaline should be that which prevailed at the time of its original synthesis in the meteorite, thus giving a clue as to the primordial enantiomeric composition of other amino acids in the meteorite. In 1975 Pollock *et al.* achieved a partial G.C. resolution of authentic D,L-isovaline as well as of the isovaline from the Murchison meteorite, obtaining comparable analytical results from each (D-III, *ca.* 52%; L-III, *ca.* 48%). This led to the conclusion that the Murchison isovaline was in fact racemic and that therefore it and the other amino acids in this meteorite had originated as racemic mixtures by abiotic, extraterrestrial syntheses. Since these conclusions were based on the demonstrated (Pollock *et al.*, 1975) non-racemization of isovaline under the ordinary laboratory conditions and did not take into account the then unknown possibility of radioracemization, the possible radioracemization susceptibility of isovaline clearly became a cosmologically pertinent question.

Our finding that solid isovaline (Table V) is approximately as susceptible to radio-racemization as are the solid common amino acids having α-hydrogen atoms (Table I) indicates that this structural difference is not pertinent as regards the racemization of solid amino acids by ionizing radiation. The probable mechanistic reason, in terms of its lack of an α-hydrogen atom, for the immunity to radioracemization of isovaline as its dissolved sodium salt, however, has been discussed above. While the enantiomeric composition of the isovaline in the Murchison meteorite was found (Pollock et al., 1975) to be *approximately* D:L/50:50, our observed radioracemization of solid isovaline suggests the need to reevaluate the earlier conclusion (Pollock et al., 1975; Lawless, 1973) that the primordial enantiomeric composition of the isovaline and other amino acids in the Murchison meteorite must therefore have been racemic. Although the time since the Murchison chondrite fragmented from its parent body is only $ca.$ 1.2×10^6 years (Cressy and Bogard, 1976), resulting in a cosmic ray radiation dose of $ca.$ 3×10^7 rads during this period (based on 10^8 rads during 3.5×10^6 years for the Orgueil meteorite (Studier et al., 1965)), the natural radioactivity of a meteorite parent body would have provided an integrated dose of $ca.$ 5×10^8 rads during the 4.5×10^9 years of its existence (Anders, 1961; Studier et al., 1965). Thus, the isovaline in the Murchison chondrite has received a total radiation dose of $ca.$ 5.3×10^8 rads, some 60% of the dose which caused 4.8% racemization of anhydrous isovaline (Table V). Although it is not presently known how the mineral matrix of the meteorite might alter the efficacy of radioracemization, it would seem probable that significant radioracemization of any non-racemic amino acids indigenous to meteorite parent bodies could have occurred during the 4.5×10^9 years since their origin. These speculations, of course, carry no implication whatsoever that the racemic amino acids found in present day meteorites were originally optically active or of biological origin. Nevertheless we believe that the phenomenon of radioracemization, in principle, makes the question of the primordial enantiomeric composition of amino acids in meteorites a fundamentally indeterminate one.

Acknowledgment

We are indebted to the National Aeronautics and Space Administration (W.A.B., N.E.B.) and to the U.S. Department of Energy (R.M.L.) for their support of portions of the above studies.

References

Anders, E.: 1961, *Ann. New York Acad. Sci.* **93**. 651.
Bada, J. L.: 1972, *Earth Planet. Sci. Lett.* **15**, 223.
Bada, J. L., Luyendyk, B. P. and Maynard, J. B.: 1970, *Science* **170**, 730.
Bada, J. L. and Protsch, R.: 1973, *Proc. Nat. Acad. Sci.* **70**, 1331.
Bada, J. L., Protsch. R. and Schroeder, R. A.: 1973, *Nature* **241**, 394.
Bernstein, W. J., Lemmon, R. M. and Calvin, M.: 1972, in *Molecular Evolution, Prebiological and Biological*, D. L. Rolfing and A. I. Oparin, eds., Plenum, New York, pp. 151–155.
Bonner, W. A.: 1973, *J. Chromatogr. Sci.* **11**, 101.
Bonner, W. A.: 1974, *J. Molec. Evol.* **4**, 23.
Bonner, W. A. and Blair, N. E.: 1979, *J. Chromatogr.* **169**, 153.
Bonner, W. A., Blair, N. E. and Lemmon, R. M.: 1979a, *J. Am. Chem. Soc.* **101**, 1079.
Bonner, W. A., Blair, N. E., Lemmon, R. M., Flores, J. J. and Pollock, G. E.: 1979b, *Geochim. Cosmochim. Acta* (In press).
Bonner, W. A. and Flores, J. J.: 1975, *Origins of Life* **6**, 187.
Bonner, W A. and Lemmon, R. M.: 1978a, *J. Molec. Evol.* **11**, 95.
Bonner, W. A. and Lemmon, R. M.: 1978b, *Bioorg. Chem.* **7**, 175.
Bonner, W. A., Lemmon, R. M. and Noyes, H. P.: 1978, *J. Org. Chem.* **43**, 522.

Bonner, W. A., Van Dort, M. A. and Flores, J. J.: 1974, *Anal. Chem* **46**, 2104.
Bonner, W. A., Van Dort, M. A. and Yearian, M. R.: 1975, *Nature*, **258**, 419.
Bonner, W. A., Van Dort, M. A., Yearian, M. R., Zeman, H. D. and Li, G. C.: 1976/77, *Israel J. Chem.* **15**, 89.
Cressy, Jr., P. J. and Bogard, D. D.: 1976, *Geochim. Cosmochim. Acta* **40**, 749.
Cronin, J. R. and Moore, C. B.: 1971, *Science* **172**, 1327.
Dungworth, G.: 1976, *Chem. Geol.* **17**, 135.
Dungworth, G., Vincken, N. J. and Schwartz, A. W.: 1974, *Comp. Biochem. Physiol.* **47B**, 391.
Evans, E. A.: 1966, *Nature* **209**, 169.
Evans, E. A., Green, R. H. and Waterfield, W. R.: 1968, in *Proceedings of the International Conference on Methods of Preparation and Storage of Labelled Compounds,* 2nd, pp. 1019–1036.
Feng, P. Y. and Tobey, S. W.: 1959, *J. Phys. Chem.* **63**, 759.
Flores, J. J. and Bonner, W. A.: 1974, *J. Mole. Evol.* **3**, 49.
Flores, J. J., Bonner, W. A. and Van Dort, M. A.: 1977, *J. Chromatogr.* **132**, 152.
Garrison, W. M.: 1968, in *Current Topics in Radiation Research*, Vol. IV, M. Ebert and A. Howard, eds., North-Holland Publishing Co., Amsterdam, pp. 45–94.
Garrison, W. M.: 1972, *Radiation Res. Rev.* **3**, 305.
Hare, P. E. and Mitterer, R. M.: 1968, *Carnegie Inst. Washington Yearbook* **67**, 205.
Helfman, P. M. and Bada, J. L.: 1975, *Proc. Nat. Acad. Sci. USA* **72**, 2891.
Helfman, P. M. and Bada, J. L.: 1976, *Nature* **262**, 279.
Kroepelin, H.: 1968, in *Advances in Organic Geochemistry, Proceedings of the Fourth International Meeting*, P. A. Schlenck, ed., Pergamon Press, Oxford, pp. 535–542.
Kvenvolden, K., Lawless, J., Pering, K., Peterson, E., Flores, J., Ponnamperuma, C., Kaplan, I. R. and Moore, C.: 1970, *Nature* **228**, 923.
Kvenvolden, K. A., Lawless, J. G. and Ponnamperuma, C.: 1971, *Proc. Nat. Acad. Sci.* **68**, 486.
Kvenvolden, K., Peterson, E. and Pollock, G. E.: 1972, *Adv. Org. Geochem.*, Proc. Int. Meeting, 5th, Hanover, Germany, Pergamon Press, Oxford-Braunschweig, pp. 387–401.
Lawless, J. G.: 1973, *Geochim. Cosmochim. Acta* **37**, 2207.
Lawless, J. G., Kvenvolden K. A., Peterson, E., Ponnamperuma, C. and Moore, C.: 1971, *Science* **173**, 626.
Lawless, J. G. and Peterson, E.: 1975, *Origins of Life* **6**, 38.
Mitterer, R. M.: 1975, *Earth Planet Sci. Lett.* **28**, 275.
Noyes, H. P., Bonner, W. A. and Tomlin, J. A.: 1977, *Origins of Life* **8**, 21.
Oró, J., Gilbert, J., Lichtenstein, H., Wikstrom, S. and Flory, D. A.: 1971a, *Nature* **230**, 105.
Oró, J., Nakaparksin, S., Lichtenstein, H. and Gil-Av, E.: 1971b *Nature* **230**, 107.
Pollock, G. E., Cheng, C. N., Cronin, S. E. and Kvenvolden K. A.: 1975, *Geochim. Cosmochim. Acta* **39**, 1571.
Schroeder, R. A. and Bada, J. L.: 1973, *Science* **182**, 479.
Studier, M. H., Hayatsu, R. and Anders, E.: 1965, *Science* **149**, 1455.
Tolbert, B. M., Krinks, M. H., Castrillan, J. A., Henderson L. E. and Finch, M. B.: 1962, Unpublished manuscript.
Ulbricht, T. L. V.: 1959, *Quart. Rev.* **13**, 48.
Ulbricht, T. L. V. and Vester, F.: 1962, *Tetrahedron* **18**, 629.
Vester, F., Ulbricht, T. L. V. and Krauch, H.: 1959, *Naturwiss.* **46**, 68.
Wehmiller, J. F. and Hare, P. E.: 1971, *Science* **173**, 907.
Wehmiller, J. F., Hare, P. E. and Kajula, G. A.: 1976, *Geochim. Cosmochim. Acta* **40**, 763.
Williams, K. M. and Smith, G. G.: 1977, *Origins of Life* **8**, 91.

INDEX

Agmenellum quadruplicatum, 72
Albedo, 101, 115-116
Algae
    blue green, *see* cyanophytes
    green, chlorophyta, 43-44
    in desert rocks, 33, 44
    in soil, 86
    mats, 90, 94, 96-97, 156
    oxygen production, 121-130
Aldan Shield, Siberia, 160-167
Amino acid, *see also* specific amino acids
    α-radical, 189-192
    abiotic origin, 192
    composition
        of halophilic enzymes, 64
        of vacuole membrane protein, 59
    enantiomer ratio, 191-192
    non-protein, 192
    radioracemization, 183-194
    transport, 65-66
2-aminobutyric acid, radiolysis and racemization, 186
Ammonia, in paleoatmosphere, 106, 107, 113
Amphibolite, 156-157, 160-162, 164
Anacystis nidulans, 72
Aphanocapsa, 77, 81
Archaebacteria, 56
Aphanothece halophytica, 70, 72, 77-81
Archean, 129, 134, 155, 161
    banded iron formations, 134, 135, 153-158
    barite, 152, 166
    cyanophytes, 137, 144-145
    igneous rocks, 154, 162
    microfossils, 134
    reducing environment, 169
    sediments, 137, 152, 160
    sulfates, 165-166
    sulfate reducing bacteria, 165
    sulfur isotope record, 149-158, 159-171
    stromatolites, 136-138
Aridity, 44
Aspartate transcarbamylase, 61
Autotrophs, 125, 126, 130, 134, 136, 138-139, 145

Bacillus caldolyticus, 47-54
Bacteria
    adaptation to high salinity, 69
    aerobic, 58
    anaerobic, 56-58
    chemotrophic, 26
    denitrifying, 106
    frozen and dehydrated, 18-21, 27-28
    green and purple sulfur, 58
    halophilic, 14, 20, 61-67
    heterotrophic, 55, 58, 71
    osmophilic, 21
    photosynthetic, 26, 55, 126, 128-129
    psychrophilic, 27-29
    psychotrophic, 27-28
    soil, 29, 33, 44
    sulfate reducing, respiring, 29, 149-157, 159-171
    sulfur oxidizing, 149, 152
    thermophilic, 26, 28, 47-54, 61
Bacteriorhodopsin, 61, 65
Banded iron formation, 134-135, 152, 155-156, 165
    Bending Lake, Canada, 153-155
    Isua, West Greenland, 128, 156-158, 162-163, 167
    Michipicoten, Canada, 153-157, 161-163, 166-167
    Schist belts of Rhodesia, 162-163, 166
    Woman River, Canada, 153-157, 162-163, 166-167
Barberton, S. Africa
    barites, 164
    greenstone belt, 164, 167
    Onverwacht group, 137
    Pongola Supergroup, 138
Barite, 152, 164, 166
Benzidinium hydrochloride, 175, 177-178
Biosynthesis, 121, 123, 125
Blind River Formation, Ontario, 97

Carbon dioxide, 123, 125-126
    in paleoatmosphere, 106-108, 112-114, 116-117
Carbon monoxide, 107, 122-123
Carbonate
    caliche, 95
    production by microorganisms, 96
    sedimentary, 139, 142, 170
Calothrix
    pulvinata, 89, 90, 94
    scopulorum, 73
Carbon isotope ratio, 135-136, 144, 156
Cell
    bouyancy, 56

Cell (continued)
  density, 56
  growth in extreme environments, 19–20, 25, 29–30, 47, 70, 73, 79–80
  permeability, 6
  surface area, 6
  survival
    of freezing, 1–23, 25
    in deep sea, 30
  swelling, 13
Cell membrane, 4, 26
Cephalophytarion grande, 96
Chasmoendolithic organisms, 35, 40
Chapman reactions, 105, 110
Chemoautotrophy, 125–126
Chemotaxis, 55
Chlorophyll a, 134
Chromatium, 152
Chroococcidiopsis, 43
Clay, and amino acid racemization, 191
Climatic change, 99, 101–102, 104
Clostridium, 56
Coelosphaerium, 78–79
Competition, 72, 124, 128
Convective energy transport, 99, 102–104
Cooling rate, 1–23
Cryptoendolithic organisms, 35–38, 41–44
Cyanobacteria, *see* cyanophytes
Cyanophytes, 35–37, 43, 55–58, 85–98, 138, 142–144, 169–170
  and atmospheric composition, 135–137
  halophily and halotolerance, 69–81
  mats, 70, 76, 85, 90, 94, 134, 138, 143
  photosynthesis, 128–129
  temperature maxima for growth, 26
Cytochrome oxidase, 63
Cytoplasm
  change in pH, 13, 66
  concentration of solutes, 13

Dactylococcopsis, 77–79, 81
Darwin, Charles, 145
Dead Sea, 74
Denitrification, 106
Desert
  climate, 33, 35, 75
  crust, 85–98
  hot and cold, 33–45
Dehydration, 44, 90, 94–95, 97
  during freezing, 13–19
Deoxyribonucleic acid, damage by ionization, 1
Desulfovibrio, 28, 149
Dew, 22, 42
Diagenesis, 153, 164, 181
  effect on amino acid racemization, 190–191
Dunaliella, 69

Entophysalis major, 76, 96
Enzyme
  active site, 50
  conformation changes with temperature, 50, 53–54
  entropy of substrate binding, 50
  salt
    inactivation, 62–63
    requirement, 61–65, 69
Enzyme kinetics
  of malate dehydrogenase, 48–51
Eoastrion, 179
Eoentophysalis belcherensis, 96
Eosynechococcus moorei, 96
Escherischia coli, 30, 64
  frozen, 3, 14, 15
Eukaryotes, 69, 168–169
  algae, 39, 44, 71
  temperature limits, 26
Evaporation, 69, 74–75, 101

Fatty acid synthetase, 62
Fermentation, 123–125
Ferredoxins, 168
Fig Tree Group, S. Africa, 152, 162–163
Fish, Arctic, 19
Flagella, 55, 58
Freezing, of cells, 1–23
Fungus
  freezing, 15, 17, 19–21
  in polar deserts, 43–44

Galaxy, 101–102
Glaciation, 101–102
Gloeocapsa nigrescens, 81, 89–90, 94–95
Glycerol
  photosynthetic production, 69
  protective effect during freezing, 8–9
Granite rock communities, 40, 44
Gravitational energy release, 99–101
Great Salt Lake, 74, 81
Greenhouse effect, 116–117
Grenville Province, Canadian Shield, 138, 162–163, 166
Gunflint Iron Formation, 134, 173

Halobacterium, 56, 58, 69, 72
  cutirubrum, 61
Halophily, 69–81
Halotolerance, 68–81
Helium, 99–100
Heterotrophy, 125–126, 130
  anaerobic, 168–169
  bacterial, 26
Humidity, 27, 33, 40, 42–44, 74
Hydrogen, 99, 100, 123, 131

Hydrogen (continued)
  electron donor in photosynthesis, 126
  species chemistry, 105–106
  volcanic origin, 122, 125–127, 130
Hydrogen sulfide, 28, 77, 150, 155–158, 160, 166, 169
  as electron donor in photosynthesis, 126
Hydrophobic interactions, and enzyme conformation, 50, 62, 63–64

Ice, intracellular, 2–7, 12–15
Ice Age, 101
Interstellar clouds, 101–104
Intertidal zone, 73, 74, 76, 79
Isocitric dehydrogenase, 61
Isovaline, radiolysis and radioracemization, 186, 188–189, 192–193
Isua, West Greenland
  amerilik basalts, 160–162, 164
  amphibolites, 160–162, 164
  banded iron formation, 162–163, 170
  sulfur isotope record, 168

Karst caves, Indiana, 28, 29
Kerogen, 135

Lakes, Arctic and Antarctic, 27
Leucine, radiolysis and radioracemization, 186–187
Leukoberbelin blue reagent, 175, 178–181
Lichen, 21, 41, 43, 90
Lipids, 47, 62
Lyngbya estuarii, 76, 96

Magnetite, 155
Malate dehydrogenase, 28
  from *Bacillus caldolyticus,* 47–54
  from pig heart, 52
Mammalian cells, 10–11, 15, 19
Manganese
  deposits
    bacterial, 175, 179
    fossil, 173
  precipitation, 173–182
Manganese dioxide, 173, 175–179
Mars, 33
Membrane, 62
  of gas vacuole, 56
  vesicles, 65
Metabolism, 121–132
Metabolic rate, 29–30, 47
  and sulfur isotope fractionation, 158
Metallogenium, 179
Metamorphism, 153, 156, 160, 166
Meteorite
  carbonaceous chondrite, 192, 193
  sulfide in trolite phase, 160–162
Methane
  biosynthesis, 125
  in paleoatmosphere, 106–108, 110, 112
Methanogen, 56, 58, 125
Methanosarcina barkeri, 56
Micrococcus radiodurans, 2
Microcoleus
  chthonoplastes, 70, 73, 76, 86, 89–90, 94–95
  vaginatus, 86, 88, 91–96
Microenvironment, 36
Microfossil, 134, 138–139
Mojave Desert, California, 44
Morphology, of manganese oxidizing bacteria, 178–181
Motility
  as a function of gas vacuole, 55–60
  of desert crust cyanophytes, 92–96
Mutation, 124, 131, 176

Natural selection, 73, 124, 131
Negev Desert, Israel, 21, 34–35, 37–38, 44
Niche, 72, 74, 80
Nicotinamide adenine dinucleotide, 48–54
NADH-menadione oxidoreductase, 62
Nitrogen
  fixation, 74
  species chemistry, 105–106, 108
Nitrous oxide, in paleoatmosphere, 106, 108
Norleucine, radiolysis and radioracemization, 186
Norvaline, radiolysis & radioracemization, 186
Nostoc
  commune, 89, 94
  endophytum, 73
Nunesuch Shale, Michigan, 152, 155

Optical activity, origin, 183, 193
Organic carbon, 122–124, 126
Organic matter, 155, *see also* organic carbon
  dissolved, 29–30
  particulate, 29–30
Oscillatoria limnetica, 72, 77
Osmosis, 61, 69
Osmotic shock, 12
Outokumpu, Finland, black shales, 162–163, 166
Oxaloacetate, 48, 51–52
Oxygen
  atmospheric, 105–119, 121–132, 137–138, 142–144, 152, 169–170
  consumption, 122, 130, 143–144
  effect on dehydration, 18
  production, 127, 129–130
  respiration, 160, 168

Oxygen (continued)
   role in banded iron formation, 128
   sink, 122, 123, 136
   tolerance, 134, 136
Ozone
   of paleoatmosphere, 105-119
   screen, 112, 138
   toxicity, 105, 110-111

Paleoatmosphere, 105-119
Paleolyngbya barghoorniana, 96
Parity violation, 183
Pedomicrobium, 179
Permeability, 6
Phanerozoic sediments, 151, 164
Phormidium, 77, 78, 79
   retzii, 89, 90
Phosphorus, as limiting nutrient, 127-128
Photochemistry, 107, 109
Photodissociation
   of oxygen, 111
   of water, 106
Photosynthesis, 27, 58, 126, 136-137, 168
   aerobic, 134
   algal, 22, 127-128, 136
   anaerobic, 77
   bacterial, 136
   cyanophytic, 170
   in desert rock communities, 36, 44
   in sulfur-oxidizing bacteria, 168-169
   oxygen release, 106, 123
Pilbara Block, Australia, 162-164, 167
Plankton, 76
   ancient, 138, 145
   bloom, 74
Plasmid, 58-59
Potassium, 65-66, 69
Precambrian, 181
   eukaryotes, 168
   organisms, 144
   prokaryotes, 96
   sediments, 135, 152, 160
   stromatolites, 74, 86
   Sudbury basin, 162-163
   sulfate reducers, 156
Precipitation, atmospheric, 101
Primary production
   Archean, 129, 137-139, 143-144
   limited by hydrogen supply, 126, 130
   prokaryotic, 95
Pressure
   as a limit of life, 25-32
   hydrostatic, 25
   of paleoatmosphere, 116, 117
Prokaryotes, 26, 43-44, 80, 85-98, 138-139,
   see also cyanophytes

   motility, 55-60
   temperature limits, 26
Proteins, *see also* enzymes
   denaturation, 63
   ribosomal, 64
   surface-to-volume ratio, 65
   thermostability, 26
Proterozoic
   atmosphere, 136-137
   carbonates, 142-143
   deposits, 135, 140-143
   microfossils, 138, 142
   primary production, 139
Protozoa, 19
Pyrite, 135, 142, 165
Pyrophosphatase, 28

Quartzite, 139

Racemization, of amino acids
   by radiolysis, 183-194
   thermal, 191
Radiation
   ionizing radiation, 1, 2
   γ-radiation, 184-185
Radiolysis, of amino acids, 183, 185-188, 189
Rain, 29, 42, 74-75
Red bed, 142-143
Red Sea, 28, 75
Rehydration, 16
Reproduction, in extreme environments, 25,
   28, 70, 71, 73, 79
Respiration
   aerobic, 122-123, 127, 130
   anaerobic, 126
Ribosomes
   effect of K^+ vs Na^+, 65
   stability and salt level, 69

Salinity, 28
   and halobacterial enzyme activity, 61-67
   change during freezing, 8-9, 20
   dependence and tolerance in cyanophytes,
   69-81
   in desert soild, 90, 95, 97
   niche, 72
Sandstone, 35, 37-39, 41-42, 44
Schizothrix
   penicillata, 89-90, 94
   subconstricta, 89-90, 94
Scytonema tenellum, 89, 90, 94
Sediments
   Black Sea, 151
   oceanic, 123, 125-129, 150
   reducing, 150
Serratia marcescens, 17-18

Sinai Desert, 35
Snow, 28, 42
Sodium, 65–66, 79
  requirement by halophiles, 69–71
Solar constant, 99–104
Solar Lake, Sinai, 74–75, 77, 79
Solar luminosity, *see* solar radiation
Solar neutrino experiment, 99, 101
Solar radiation, 39, 99–104, 115–116
Solar spectrum, 104, 107
Solar system, 99–104
Solar wind, 101–102
Solar zenith angle, 111–114
Solution-effect injury, 8–9
Sonoran Desert, Mexico, 39–40, 44
Southern Victoria Land, Antarctica, 27, 33–35, 39, 41–44
Speciation, and salinity gradient, 70, 79–80
Spirulina
  subsalsa, 76, 78–79
  tenerrima, 76
Stromatolites, 74, 86, 96, 134–139, 142–144
Subterranean deposits, 28
Sulfate, 126, 150, 164–165, 169
  from weathering of sulfide, 152
  plant metabolism, 156
  reduction, 29, 149–158, 159
  respiration, 168
Sulfide, 164–165
  Adirondack, 167
  Balmat-Edwards, 167
  from bacterial sulfate reduction, 150–151, 155
  meteoritic, isotopic composition, 150, 160–162
  in Precambrian sediments, 152, 156
  Sudbury District, 162–163, 167
Sulfur
  bacteria, *see* bacteria
  cycle, 149, 161
  isotope fractionation, 149–158, 160–168
  magmatogenic, 164, 166, 169
  sedimentary, 164
Supernova, 102
Synechoccus, 77–79, 81
Synechocystis, 77–79, 81
Synthesis, abiotic, 123–125

Temperature
  extremes, 25–32, 33, 39–44
  high, 47–54
  low, 1–23
  of paleoatmosphere, 105, 109, 114–116
  of terrestrial surface, 105, 114, 116
Tetramitus, 78–79
Thawing rate, 7, 10–12, 19
Thermal tolerance, 47
Thermonuclear reactions of sun, 99–104
Threonine deaminase, 62
Tindaria callistiformis, 25
Trichodesmium, 74, 138
  thiebautti, 78–79

Ultraviolet radiation, 101, 105–119
Uraninite, 135, 142

Vacuole, organelle of motility, 55–60
Valine, radiolysis and radioracemization, 186
Vertical eddy diffusion, 110
Virus, 26
Vitamins, 29
Volcanoes, 26, 123, 153
  as source of atmospheric gases, 122, 125–127, 130, 155–156

Water
  absorption of solar radiation, 112, 114, 116–117
  activity, 1–23
  bound, 14–15, 17
  content of cells, 1–23
  electron donor in photosynthesis, 127
  liquid, 21–22
  supercooled, 5
  vapor, 21–22, 106, 115
  vapor mixing ratio, 109

Xenococcus, 77–79
Xeromyces bisporus, 21

Yallahs ponds, Jamaica, 74–75, 77, 79
Yilgarn Block, Australia, 160–163, 166
Yeast
  freezing, 6–7, 10–16
  in soils, 33

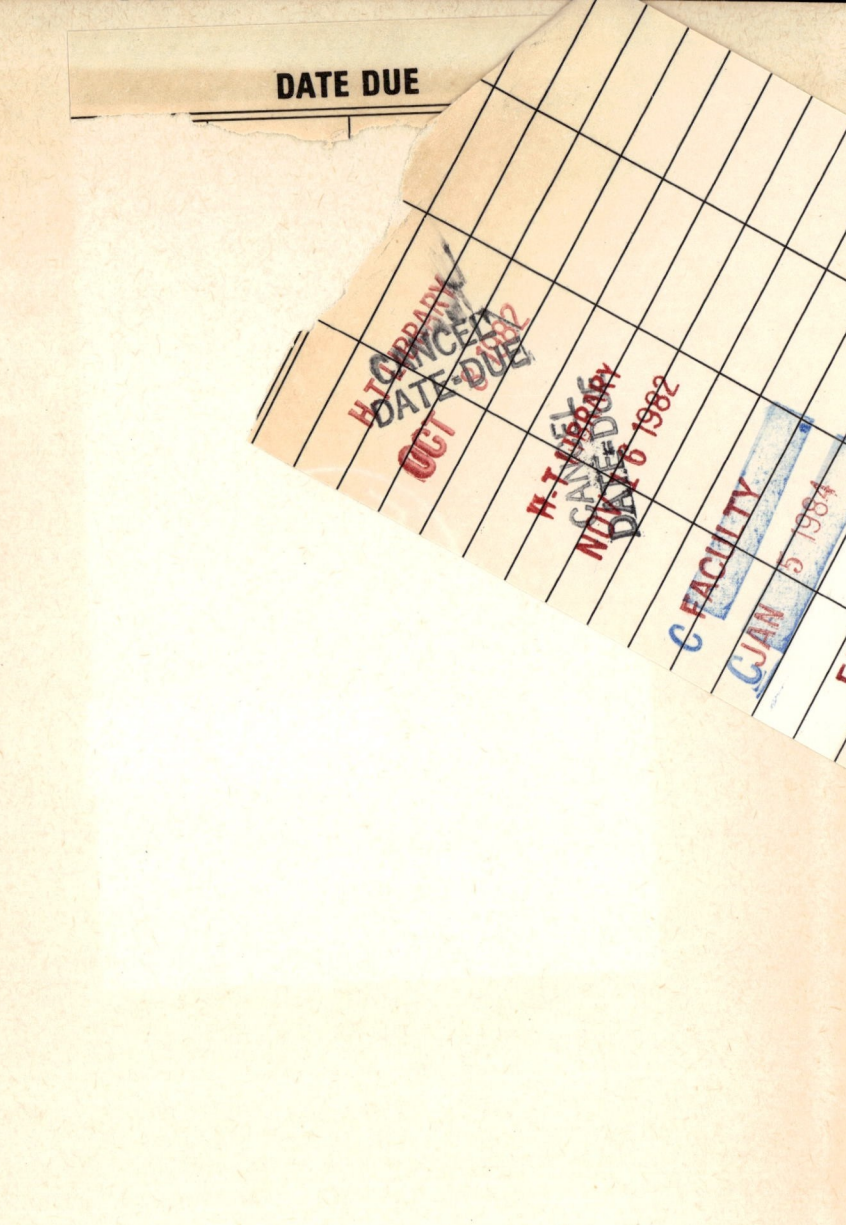